S0-BQM-345

THE CYTOPATHOLOGY OF ESOPHAGEAL CARCINOMA

Precancerous Lesions and Early Cancer

Yi-Jing Shu, M.D., M.I.A.C.

Deputy Head of Cytopathologic Department
Cancer Research Institute
The Chinese Academy of Medical Sciences
Peking, China

Visiting Professor, Department of Pathology
School of Medicine
University of Colorado Health Sciences Center
Denver, Colorado

Visiting Professor, Department of Pathology
Albert Einstein College of Medicine
Montefiore Hospital and Medical Center
New York, New York

Edited and adapted by

Leopold G. Koss

Professor and Chairman
Department of Pathology
Montefiore Hospital
Bronx, New York

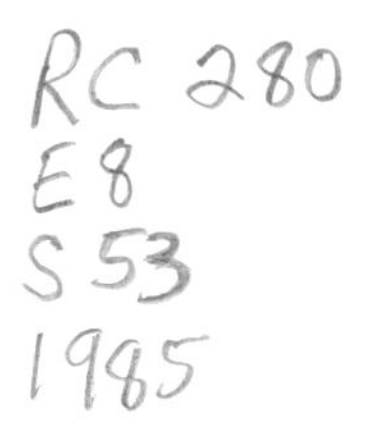

MASSON Publishing USA, Inc.
New York • Paris • Barcelona • Milan • Mexico City • Rio de Janeiro

Library of Congress Cataloging in Publication Data

Shu, Yi-Jing.
Cytopathology of esophageal carcinoma.

Includes bibliographical references and index.
1. Esophagus—Cancer—Diagnosis. 2. Diagnosis, Cytologic. 3. Exfoliative cytology. 4. Esophagus—Cancer—China. I. Koss, Leopold G. II. Title.
[DNLM: 1. Esophageal Neoplasms—pathology. 2. Carcinoma, Squamous Cell—pathology. 3. Precancerous Conditions. 4. Cytodiagnosis. W1 MA9309RN v.5 / WI 250 S562c]
RC280.E8S53 1985 616.99′43207582 84-5669
ISBN 0-89352-171-X

ISBN 0-89352-171-X
Library of Congress Catalog Card Number: 84–5669

Printed in the United States of America

Distributed by Year Book Medical Publishers

Masson Monographs in Diagnostic Cytopathology

Series Editor: William W. Johnston, M.D.

1. *Diagnostic Respiratory Cytopathology*

By William W. Johnston, M.D.
and
William J. Frable, M.D. (1979)

2. *Aspiration Biopsy for the Community Hospital*

By David B. Kaminsky, M.D. (1981)

3. *Cytopathology of the Central Nervous System*

By Sandra H. Bigner, M.D.
and
William W. Johnston, M.D. (1983)

4. *Cytopathologic Interpretation of Transthoracic Fine-Needle Biopsies*

By Thomas A. Bonfiglio, M.D. (1983)

5. *The Cytopathology of Esophageal Carcinoma: Precancerous Lesions and Early Cancer*

By Yi-Jing Shu, M.D.
Edited and adapted by
Leopold G. Koss, M.D. (1985)

Foreword

In 1943, Papanicolaou and Traut published their now historical book, *The Diagnosis of Uterine Cancer by the Vaginal Smear,* heralding the onset of a new era of cancer detection by cytologic approaches. Subsequent events have shown that, indeed, the simple smear of the uterine cervix, now commonly referred to as the Pap smear, is a powerful tool of prevention of invasive carcinoma of the uterine cervix.

The application of the cytologic technique to the detection and early diagnosis of cancers of other organs followed rapidly. Among others, endometrial carcinoma, lung cancer, and cancer of the urinary tract and of the gastrointestinal tract became the subject of these inquiries, none of them approaching in scope or results the effectiveness of cytologic screening for cervical cancer.

Carcinoma of the esophagus has received only scanty attention in the United States because of the relative rarity of the disease. The first major paper on this subject, which appeared in 1949, was by Andersen, McDonald, and Olsen from the Mayo Clinic. I recall vividly early work on the same subject that took place in my laboratory during the early 1950s and resulted in a report published in *Cancer* in 1955 on 364 patients with washings of the esophagus. Dr. Papanicolaou participated in this work, and the illustrations were prepared by his personal photographer, Mr. Constantine Reily.

The thought that esophageal cytology could sometimes lead to the diagnosis of unsuspected esophageal carcinoma *in situ* was first expressed in the 1961 edition of my book, *Diagnostic Cytology,* from which I quote: "It is more than likely that all of the epidermoid cancers of the esophagus are preceded by an *in situ* stage during which the cancer is still limited to the mucosa and presumably curable. Unfortunately, the study of this problem is still ahead of us and will be in all likelihood a formidable undertaking." At that time I knew little about the fact that there are several areas in the world where carcinoma of the esophagus is a common disorder, claiming the lives of numerous, sometimes young victims. Thus, on the littoral of the Caspian Sea, in some areas of Southern Africa, in the Brittany region of France, to name a few, esophageal cancer is endemic.

For many years after the 1949 revolution, China remained virtually closed to all scientific inquiries. Thus it was a major surprise to me to hear during the 1974 International Cancer Congress in Florence, Italy, about the work on cytologic detection of esophageal carcinoma in Henan Province. As the restrictions became relaxed, the people behind this remarkable effort became known to us. Thus, it was a signal pleasure to welcome to our Department the author of this book, Dr. Yi-Jing Shu, from the Cancer Research Institute of the Chinese Academy of Medical Sciences in Beijing.

Dr. Shu spent a year at Montefiore Hospital as a Visiting Professor of Pathology at the Albert Einstein College of Medicine. There were many occasions during this period to learn from Dr. Shu about the truly unique effort to identify and prevent a mortal disease affecting a large population in China. The ingenuity with which this program has been implemented, the perseverance shown by the medical personnel and by the patients, and the remarkable results which have truly revolutionized the prognosis of carcinoma of the esophagus, will clearly belong to major accomplishments in preventive oncology. As an incidental benefit, the sequence of morphologic events in the genesis of epidermoid carcinoma of the esophagus has been at least in part unravelled. The similarities between the events in carcinogenesis in the esophagus and in the uterine cervix are remarkable.

Dr. Shu wrote this book in part during his tenure with us and I followed his progress with great personal interest. His is a remarkable account of a conquest of an important cancer and the credit for it must go not only to the author but also to the many people who performed the endless tasks associated with it. Dr. Shu's book is not only a highly interesting scientific account but also a monument to preventive oncology and to the skills and accomplishments of the Chinese scientists.

Leopold G. Koss, M.D.
Professor and Chairman
Department of Pathology
Montefiore Hospital
Bronx, New York

Foreword

This monograph offers, to those who do not read Chinese, the opportunity to learn from the extensive experience of Dr. Shu and his colleagues in the study of esophageal carcinoma. This tumor is relatively infrequent in our society. It does, however, occur often enough that we must be aware of its existence and know more about its early detection and management. Although the emphasis is on cytopathology, this book goes beyond the technical details and offers to those interested in epidemiology of cancer and geographic pathology the opportunity to enlarge their concepts. The relationship of dysplasia to *in situ* and invasive cancer is evaluated and the results are similar to those seen among those entities in the uterine cervix. Clues to the etiology of the esophageal cancer are offered, and it is particularly informative to those of us interested in experimental and environmental carcinogenesis that the chicken in China is also at high risk for esophageal cancer.

We have profited from Dr. Shu's visit, and this volume offers to many others the results of his studies of esophageal carcinoma.

G. Barry Pierce, M.D.
Professor and Chairman
Department of Pathology
University of Colorado

Robert H. Fennell, Jr., M.D.
Professor of Pathology
University of Colorado

Preface

China has the highest mortality rate from esophageal cancer in the world; it seriously threatens the life and health of the people. Statistics reveal that in most patients seen in the city hospitals, the disease was already in the advanced stage. In an attempt to control the disease by early diagnosis, the balloon technique was developed approximately 20 years ago. It is simple, accurate, easily accepted by the patients, and fit for mass survey and daily outpatient work. It has proved to be the best method of early detection of esophageal cancer. It is now widely used in China and is accepted as a diagnostic method by WHO. Presently, esophageal cytological examination can be used, not only in the diagnosis of esophageal cancer, but also in the detection of dysplasia. It has been used as a research tool to study the relationship between epithelial dysplasia and cancer and its natural history. Utilization of this technique has made possible epidemiologic studies, thereby coordinating etiologic research in esophageal cancer.

To date, there have been few reports of our work in languages other than Chinese. While in China, I had much encouragement from visitors abroad for such an endeavor. The materials that I have quoted in this monograph are the results of studies conducted by my colleagues in China, and to them I owe my deepest thanks. Special thanks to Dr. Qiong Shen and Song-liang Qiu. I am greatly indebted for the generous support of the Department of Pathology, University of Colorado Health Sciences Center; Albert Einstein College of Medicine at Montefiore Hospital; and in particular, Dr. G. Barry Pierce's and Dr. Leopold Koss's special interest in fostering international exchange and understanding of this monograph. I wish to give special thanks to Dr. Koss for his support and many corrections; Dr. Robert H. Fennell, Jr. for his editorial advice, encouragement, and many helpful discussions; and Dr. Henry Chu for his generous help in earnest correction of the English in the manuscript. I also wish to thank Mrs. Frances Freck for her kind help in proofreading the manuscript; the secretarial staff for their patience; and especially LaVonne King and Luann Bergquist for typing my manuscript.

This monograph is directed not only to the pathologist and cytotechnologist, but also to the oncologist and the clinician who must care for patients and decide the diagnostic approach and treatment.

Yi-jing Shu, M.D., M.I.A.C.

Contents

CHAPTER 1

The General Situation of Esophageal Cancer in China

INTRODUCTION

Cancer of the esophagus is one of the most important neoplastic diseases in China. China has the highest recorded mortality rate from this disease in the world. The age-adjusted mortality rate per 100,000 was calculated in 1970 to be 31.66 for males and 15.93 for females.[1] Deaths from this disease account for 26.5% of all cancer deaths in males and 19.7% in females. (Table 1-1).[2] In 1977, 156,876 cases of esophageal carcinoma were recorded.[2] Cancer of the esophagus is second to gastric cancer in frequency among males, and third among females, after gastric cancer and cancer of the uterine cervix. The incidence of esophageal cancer in China varies according to the geographic area. Linxian County, situated in Henan Province, probably has the highest annual rate in the world—161.33 per 100,000 for males and 102.88 per 100,000 for females.

TABLE 1-1

Age-Adjusted Death Rates per 100,000[a] for Selected Cancers by Sex and Percent Distribution of Death Rates by Site, China 1973–1975[b]

	Male		Female	
Cancer	Rate	Percent	Rate	Percent
Stomach	32.36	27.1	15.93	19.7
Esophagus	31.66	26.5	15.93	19.7
Cervix uteri	—	—	14.60	18.1
Liver	19.96	16.7	8.07	10.0
Lung	10.25	8.6	4.75	5.9
Colon/rectum	6.35	5.3	4.71	5.8
Nasopharynx	3.40	2.8	1.77	2.2
Breast	—	—	3.80	4.7
Leukemia	2.87	2.4	2.30	2.9
Others	12.75	10.6	8.80	11.0
Total	119.60	100.00	80.70	100.00

[a] Age-adjusted to world standard population, 1960.

[b] Data from the National Cancer Control Office of the People's Republic of China.

Ancient Chinese medical literature contains many descriptions of the symptoms, causes, and treatment of esophageal cancer. In *Huang Di Nei Jing Su Wen,* China's earliest medical work, written about 300 B.C., the symptoms are described as "difficulty in swallowing, blockage of the gullet, and vomiting after eating." Though this description may include gastric cancer, it documents that obstructive diseases of the upper digestive tract were already common in ancient times.[3,4]

EPIDEMIOLOGIC FEATURES[2,17]

Features of Geographic Distribution

Usually the high-incidence areas show irregular, concentric, belt-like distributions. Areas in China with high rates of esophageal cancer include:

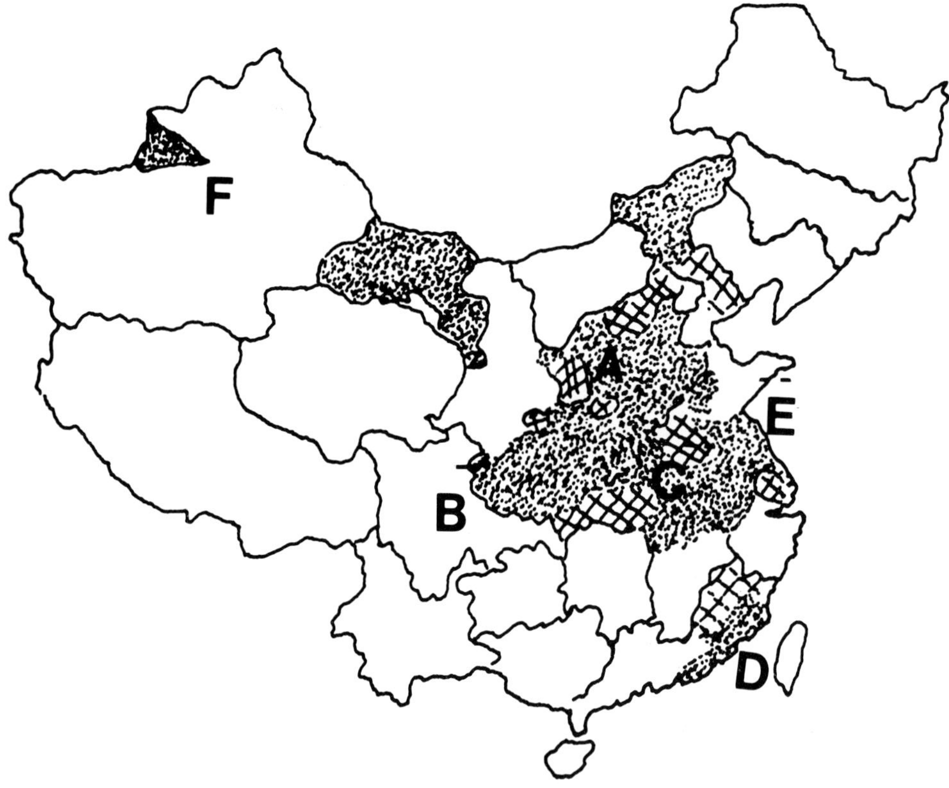

Fig. 1-1. High mortality rate areas of esophageal cancer are mainly concentrated at the Tai-heng mountain in northern China. This map shows high mortality areas (cross-hatched) and higher (dotted).

A. The Tai-heng mountain area at the border of Henan, Hebei, and Shanxi Provinces in north China.
B. Northern Sichuan.
C. Da-bie mountain area on the border of Anhui and Hubei.
D. South Fukien and regions of north-eastern Guangdong.
E. Jiangsu's northern area.
F. Northern Sinjiang area occupied by the Kazak nationality.

In these areas, mortality rates decline from the foci of high rates, producing irregularly concentric belts with declining rates (Fig. 1-1).

Stability of Rates

In several areas of China, *studies show little fluctuation in incidence and mortality rates over time* (Fig. 1-2). Age-specific esophageal cancer mortality rates have also been stable.[10]

Fluctuation in Sex Ratios by Rate

Esophageal cancer in China is more common in males (M/F:2/1). *In high rate areas the sex ratio is often close, whereas in low rate areas the opposite is true. Of particular note is the Mai County area of Guangdong,* in which the sex ratio is only 0.63 (M/F), though here unusually high overall rates of esophagus cancer are not typical.

Differences by Nationality

The mortality rate of esophageal cancer is 2–43 times higher among the Kazak in Sinjiang, than among other minority nationalities in China (Table 1-2). The age-specific

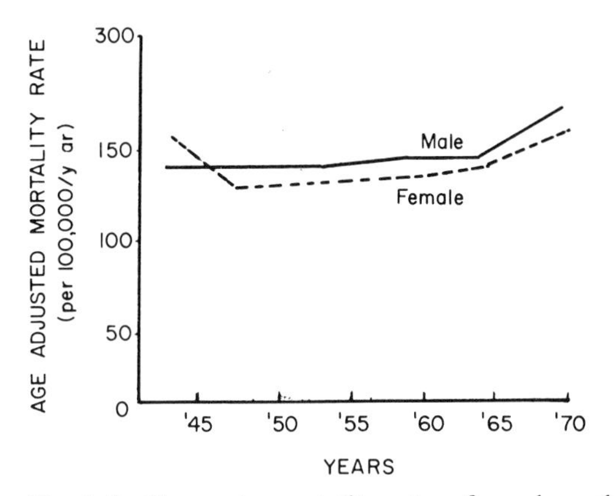

Fig. 1-2. Change in mortality rates of esophageal cancer in 76 brigades of Linhsien 1941–1970. (From The National Cancer Control Office of The People's Republic of China: *The Investigation of Cancer Mortality in China* (in Chinese). The People's Health Publishing House, Peking, 1980.)

mortality rate curves show the same configuration. *Although they live in close proximity to each other,* the Kazak, Uygur, and Mongol nationalities display large differences in esophageal cancer mortality rates.

Gullet Cancer in Chickens

Studies of several areas in China reveal that those in which esophageal cancers in humans are prevalent also exhibit a high incidence of gullet cancer in chickens. This animal tumor shows clinical and pathologic features comparable to the human disease. The relationship between the two conditions as suggested by their high common frequency in certain geographic areas can be seen in Table 1-3. As an interesting aside, chickens raised by migrants from high rate areas also have high rates of gullet cancer, often much higher than do chickens raised by the native population.[5-7]

TABLE 1-2
Age-Adjusted Mortality Rates[a] of Esophageal Cancer by Minority Nationality in China[b]

	Rates (per 100,000)	
Nationality	Male	Female
Kazak	39.27	27.08
Hui	18.90	6.32
Mongol	12.89	5.73
Uygur	12.87	7.93
Zang	7.80	5.34
Korean	5.82	1.62
Yi	1.67	0.91
Miao	1.61	0.63

[a] Age-adjusted to China's population, 1964.
[b] Data from the National Cancer Control Office of the People's Republic of China.

HIGH-INCIDENCE AREAS OF TAI-HENG MOUNTAINS[2]

High-incidence Areas at the Juncture of the Three North China Provinces

The high-incidence regions for cancer of the esophagus in north China are the southern Tai-heng Mountains where Henan, Hebei, and Shanxi Provinces meet, including Linxian, Anyang, Tangyin, Qixian, Huixian Counties, and Hebi City in northwestern Henan Province; Shexian, Wu'an, and Cixian Counties in southwestern Hebei Province; and Yangcheng, Jincheng, Qinohui, and Gaoping Counties in southeastern Shanxi Province. The mortality rate decreases outwardly around the high-incidence counties, which can be traced to form an irregular concentric circle (Fig. 1-3).

Northwestern Henan Province

Henan Province has the highest incidence and mortality rate of esophageal cancer in China (Fig. 1-4). The 1973–1975 statistics show that the mortality rate for the province adjusted to the whole-nation population is 32.22 (43.55 for males and 22.47 for females), comprising 40.55% of the overall deaths caused by malignant tumors. About 173 persons die from various malignant tumors every day; 70 of them are victims of esophageal cancer. Anyang prefecture in the northeastern part of the province suffers as high a mortality rate of the disease as 73.98. With a vast plain to the east, this prefecture borders on Shanxi Province on the west, faces Shandong Province across the Huanghe

TABLE 1-3

Comparative Survey of the Prevalence Rate of Pharyngeal and Esophageal Carcinomas in Domestic Fowls From Linxian and Fanxian, Henan Province[a]

	Linxian			Fanxian		
Age (years)[b]	No. of chickens	No. of carcinomas	Prevalence rate (per 100,000)	No. of chickens	No. of carcinomas	Prevalence rate (per 100,000)
0	3460	0	0	3911	0	0
2	11,563	6	51	6087	2	32
5	2617	20	760	518	0	0
7	941	6	630	62	0	0
10	193	1	510	10	0	0
Total	18,774	33	175.78	11,399[b]	2	17.55

[a] Data from the National Cancer Control Office of the People's Republic of China.
[b] Age of 819 chickens uncertain.

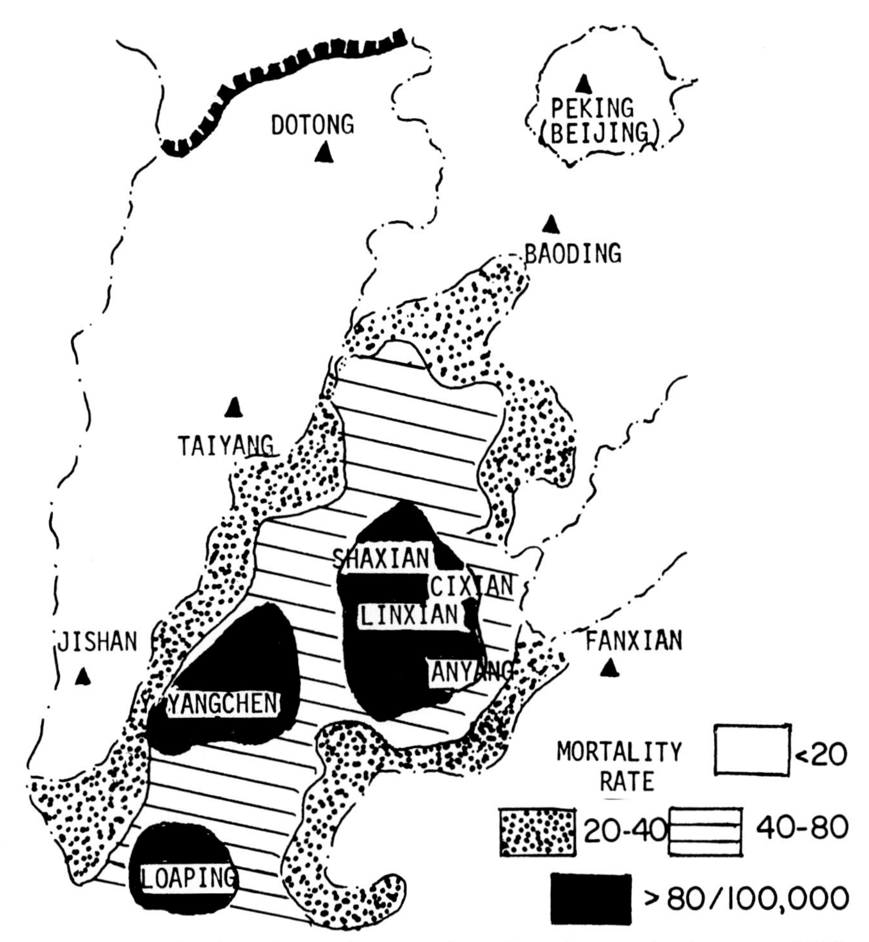

Fig. 1-3. Distribution of mortality rate of esophageal cancer in the areas of Taiheng mountain. It shows irregular concentric belt-like distribution around Taiheng mountain. (From The National Cancer Control Office of The People's Republic of China: *The Investigation of Cancer Mortality in China* (in Chinese) The People's Health Publishing House, Peking, 1980.)

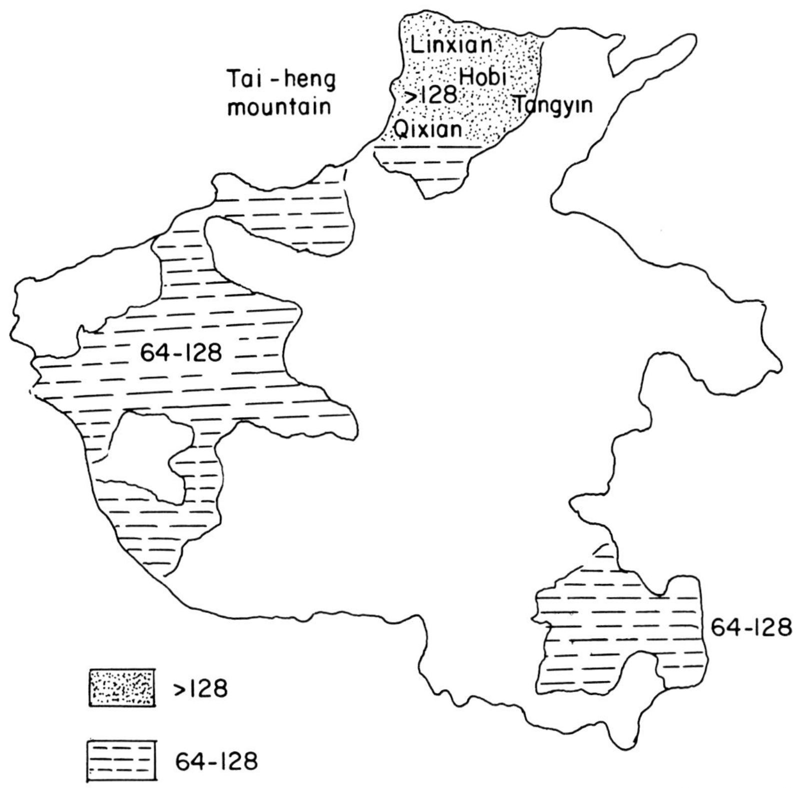

Fig. 1-4. Distribution of the mortality rate of esophageal cancer in males of Henan province, which has the highest incidence and mortality rate of esophageal cancer in China.

River on the southeast, and neighbors Hebei Province on the north. There are 12 counties and two cities with the Wei River flowing from south to northeast to empty into the Grand Canal and bisecting the prefecture into eastern and western parts, each with seven county-level administrative districts (Fig. 1-5). The western part is largely mountainous and hilly, with high incidence and mortality rates of esophageal cancer. The counties with a male mortality rate of over 128 are Linxian (161.33), Tangyin (147.98), Qixian (131.91), and Hebi City (128.37). The male mortality rate of Anyang County (122.04) is close to the figures above.

The mortality rate gradually decreases as one moves eastward, with the Wei River as a line of demarcation. The yearly average mortality rate on the plain to the east of the river falls markedly and the mortality rates on the two banks of the river are also different. The average mortality rate in the areas west of the river is 106.76, contrasting sharply with the rate of 42.75 on the opposite bank. Fanxian County to the extreme east of the river has a rate of only 23.87. Such a geographical distribution is similar to that seen in the Caspian littoral of Iran, where the mortality rate of esophageal cancer decreases from east to west.[17]

Linxian County

In Linxian County (located in the western part of Anyang Prefecture), deaths from esophageal cancer have been recorded since 1959. Recent statistics show that the adjusted mortality in 1973–1975 is 133.07 (166.33 for males and 102.88 for females), the highest of all counties in the country. A random examination of deaths of people over 30 years old in 76 production brigades reveals that the

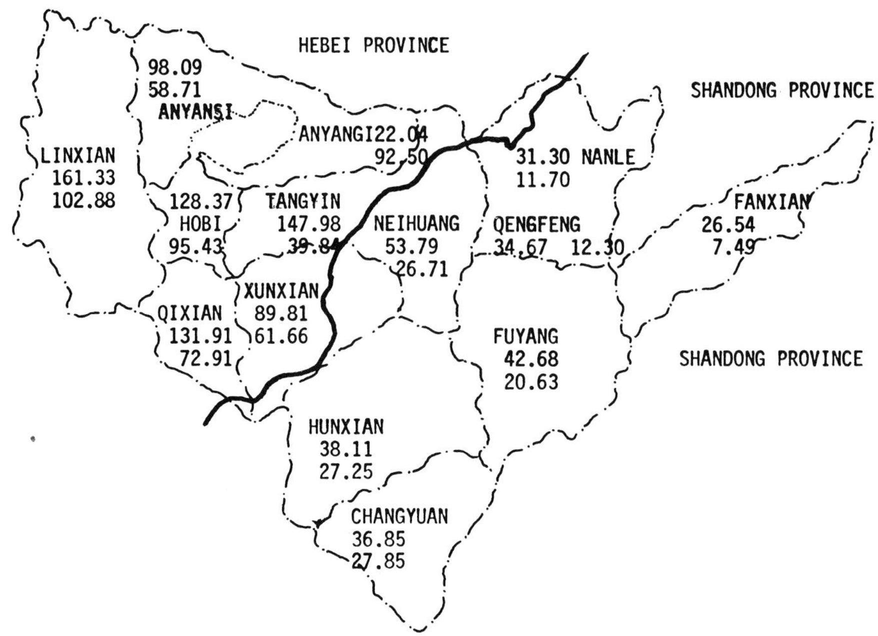

Fig. 1-5. Anyang administrative region, in the northeastern part of Henan province, suffers the highest mortality rate of esophageal cancer. There are 12 counties and two cities in Anyang prefecture; the higher incidence and mortality rates of esophageal cancer occur in the counties west of the Wei River (heavy line) where the terrain is largely mountainous and hilly. Age-adjusted mortality rates of esophageal cancer for the male and female population per 100,000 (1971–1974). (From the National Cancer Control Office of The People's Republic of China: *The Investigation of Cancer Mortality in China* (in Chinese). The People's Health Publishing House, Peking, 1980.)

adjusted mortality rate in 1941–1970 is 130.3 (134.83 for males and 126.15 for females) (Fig. 1-4). Mortality rises markedly from the over-30-year group and increases with advancing age (Fig. 1-6), causing the 60–69 age group to be the most severely affected in terms of proportional ratio (37–39%). The next two age groups are the over-70 and 50- to 59-year groups with respective proportional ratios of 28 and 23%. Over 60% of the esophageal cancer deaths occur among those between 50 and 69 years with the male age-specific mortality rate higher than the female, and about 5 years less than the average mortality rate of all age groups in the whole of Henan Province. Though the incidence and mortality fluctuate from year to year, the figure remained within 100–150 per 100,000 population, more or less constant

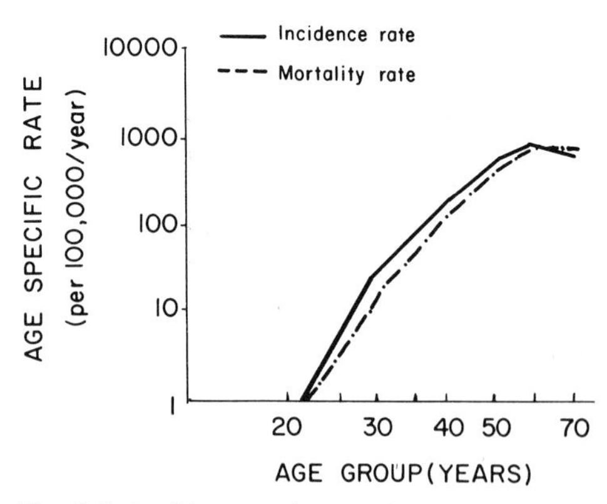

Fig. 1-6. Incidence and mortality rates (1959–1970) in Linxian county, the western part of Anyang prefecture. Mortality there is the highest of all counties in China. (From the National Cancer Control Office of The People's Republic of China: *The Investigation of Cancer Mortality in China* (in Chinese). The People's Health Publishing House, Peking, 1980.)

over a 30-year period (1941–1970). Change in various age groups is not marked (Coordinating Group for Research on Etiology of Esophageal Cancer in North China, 1975). The average mortality rate in the northern part of Linxian County is 131.94, higher than the 79.88 rate in the southern part. The male–female mortality ratio is 0.97:1 in the northern part, 1.30:1 in the southern part. In the seven northern communes, Yaocun, Rencun, and Shibanyan have higher mortality rates. Figure 1-7 is based on a study of 1959–1977 statistics and indicates the mortality rates in the county's 15 communes.

There is a sharp contrast in the mortality

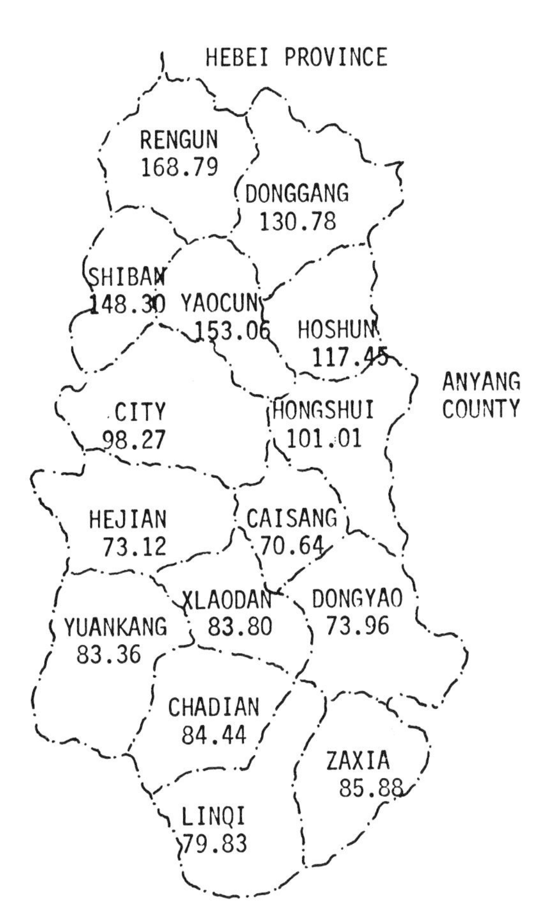

Fig. 1-7. Mortality rates per 100,000 of esophageal cancer in Linxian county by communes (1959–1977). (From the National Cancer Control Office of The People's Republic of China: *The Investigation of Cancer Mortality in China* (in Chinese). The People's Health Publishing House, Peking, 1980.)

distribution with reference to different areas. In addition to the counties of Linxian and Anyang, high-incidence centers in the northwestern Tai-heng Mountains, the Henan Cancer Prevention and Control Office reported in 1977 that there is another high-incidence center around Songxian, Iuoning, and Luanchuan Counties surrounded by the Qinling Mountains in the western part of the province. The male age-adjusted mortality rates of the three counties in 1974–1976 are 123.45, 114.31 and 105.09, respectively, second only to Anyang prefecture. There is still another relatively high-incidence area in Dabie Mountains in the eastern part of the province, centering around Huangchuan, Guangshan, and Gushi Counties (male mortality 76.60, 68.78, and 65.35, respectively), a little lower than that in the northwestern and western mountainous areas. Considering the geographical characteristics of the province, one will notice that the incidence of esophageal cancer has a close relationship to terrain; in other words, the mountain area has a higher rate than the hilly area, which in turn surpasses that on the plain; the plain near the mountains has a higher rate than the plain on the lower reaches of the Huangne River.

RISK FACTORS

The etiology of esophageal cancer is still uncertain. Recent epidemiologic and laboratory studies have implicated the role of the natural environment and life-styles of persons in high-risk areas. At present, preliminary indications suggest the following factors in development of esophageal cancer[8]:

Fermented and Moldy Foods—Pickled Vegetables

The high-rate areas of esophageal cancer in China tend to be dry, arid, and lacking foliage. Water tends to be in short supply and droughts are common. Vegetables are difficult to grow and those available are often preserved through pickling. Ingestion of pickles has become an ingrained habit.

1. Research indicates a correlation between consumption of pickled vegetables in an area and the local rates of esophageal cancer. Mortality rates are highest among those who have ingested the largest amount of pickles for the longest period of time.
2. Ames' test results show mutagens in the pickled vegetables.
3. The fungus *Geotrichum candidum link* was present as a contaminant in 83% of pickles sampled; there was additional contamination by *Aspergillus flavus* and *Fusarium moniliforme*. In laboratory studies, these fungi enhanced the development of epithelial cancers in mice administered nitrosamines.
4. In long-term studies in Linxian County, extracts and soup made with pickled vegetables produced dysplasia of the esophagus and gastric, cardiac, and liver tumors in rats (two cases). Esophageal gastric dysplasia and adenocarcinoma of the stomach developed in another rat after exposure. Liver tumors developed in two other rats and an endothelial tumor of the chest developed in yet another rat. Controls were free of tumor.[9]

Other Fermented and Moldy Foods

Studies in China reveal that in some esophageal cancer areas, ingestion of fermented pickles is uncommon. However, ingestion of other fermented and moldy food is common, and contamination of foodstuffs is severe. In certain high-risk areas in northern China, residents commonly eat breads baked more than 3–5 days previously that are overgrown with molds. The Kazaks in Sinjia, who have high esophageal cancer rates, frequently ingest yogurt and cheese contaminated with molds.

1. Among the Kazaks, Uggurs, and Mongols residing in the same area, the high-risk Kazaks had greatest level of mold contamination in their food ($p < 0.001$). Of particular interest is the differences in the types of cheese frequently ingested by these groups. The Kazaks' cheeses are large in size, more moist, and thus frequently contaminated with mold.
2. Using methylbenzylnitrosamine (MBNA) to induce animal tumors, the addition of moldy corn bread increased the rate of cancer development and decreased time to cancer. The findings show enhancement of nitrosamine carcinogenesis by molds.
3. When moldy corn bread was fed to 16 rats, three developed esophageal cancer 454–669 days later. Further testing resulted in one case of esophageal cancer, two cases of forestomach cancer and papilloma in forestomach and esophagus. In the control group fed uncontaminated corn bread, the papillomas were observed but no cancers developed.

Nitrosamine and Precursors

1. Comparisons show higher levels of nitrosamines in areas with high esophageal cancer rates. Among foodstuffs, pickled vegetables, corn, and bran of millet were found to contain dimethylnitrosamine (DMNA), diethylnitrosamine (DENA), and methylphenylnitrosamine (MPNA).
2. *Formation of nitrosamine by metabolic conversion by fungi*: Collaboration between epidemiologists and laboratory scientists in China has provided evidence that nitrosamine formation from precursors can be produced by metabolic action of certain fungi.[11] Experimental investigation demonstrated that a new volatile nitrosamine, *N*-1-methylacetonyl-*N*-3-methyl-butylnitrosamine (MAMBNA) can be synthesized by some species of fungi such as *Fusarium moniliforme, Aspergillus flavus,* and *Geotrichum candidum.* MAMBNA was shown to be positive in the Ames' test

and could induce papilloma of the forestomach and esophagus and breast adenomas.

Trace Elements

Studies of high rate esophageal cancer areas in China show relative deficiency of molybdenum, copper, cobalt, zinc, manganese, and iron in drinking water; molybdenum, nickel, manganese, titanium, and iron in food. Examination of serum, hair, and urine shows that levels of molybdenum, zinc, and magnesium are lower in high-risk areas than in low-risk areas.[12] In the high cancer areas, deficiencies of Mo and Mn in the soil and water are particularly evident. The deficiencies may promote formation of nitrosamines from precursors, particularly through the action of fungal contaminants. Animal studies show that addition of ammonium molybdate can decrease gastric adenocarcinomas associated with exposure to *N*-nitroso compounds ($p < 0.05$). Further studies are needed about the relationship between trace metals and esophageal cancer development.

Nutrition Imbalance

Studies in Sichuan, China show that the amount of fresh vegetables in the diet is inversely correlated with esophageal cancer rates ($r = -0.87$, $p < 0.01$). Comparison of the data for Kazak, Uygur, and Mongol minorities in Sinjiang shows deficiency of fruits and vegetables in the high-risk Kazak. In Hubei, studies show that migrants eat more flour, corn, and sweet potato, whereas the indigenous people eat rice and more vegetables, beans, fish, meat, eggs, and oil. These migrants continue to have high cancer rates. Urinary studies show deficiencies of vitamins B_1 and C and calcium among the migrants. Studies in Linxian (Henan) show comparable results.[17]

Chronic Esophagitis

Cytologic studies of the esophagus in Linxian showed that 14.7% of 197 patients with dysphagia had chronic esophagitis. This condition is more common in the high-rate areas for cancer. Oral leukoderma is also more common in the high-rate area of Linxian as compared with the low-rate area, Shen Yang. Case-control studies show that the risk of esophageal cancer is four-fold higher in persons with oral leukoderma.

Other Etiologic Factors

Studies in China and elsewhere suggest the etiologic role of alcoholic beverages, tobacco, tea, and genetic factors. Coarse, hard, hot foods eaten rapidly in a stooping position may also be factors, though our studies have not yet confirmed this.

CHAPTER 2

An Overview of Esophageal Cytopathology in China

In 1963, Shen and associates[13] reported their success in collecting cytological specimens from the esophagus using the especially designed inflatable balloon covered with a mesh net. In 1975, Shu and associates[14] developed a smaller balloon that was more easily swallowed and more readily accepted than the previous type. Thus, as the collecting technique improved, the use of cytopathology for the diagnosis of esophageal cancer advanced.

The esophageal cytological examination can be used not only in the diagnosis of esophageal cancer but also in the detection of other pathological changes, such as inflammation and epithelial dysplasia of the mucosa. In recent years, since the balloon technique is simple, painless, and easily repeated, it has been widely utilized to collect many specimens. A cytological examination has been used as a research method to study the relationship between epithelial dysplasia and cancer of the esophagus. Therefore, it is possible to study a progression in the development of esophageal lesions. We also investigated the components of smears as a research method for coordination of epidemiology and etiology. From 1975 to 1979, more than 500,000 persons have been examined. The range of application has been introduced as follows.

ROUTINE EXAMINATION FOR DIFFERENTIATION BETWEEN BENIGN AND MALIGNANT LESIONS IN THE ESOPHAGUS

Before 1963, esophageal and gastric cancers were diagnosed by endoscopy. Since then, symptomatic patients have been examined routinely by using the balloon technique now in general use throughout China. The detection rate of esophageal cancer ranges from 87.8 to 99% (see Table 2-1).

EARLY DETECTION

Individuals over 30 years of age were selected for mass screening surveys. In these surveys over 80% of the defined population in the area were examined. Between 56 and 85% (average 74%) of the lesions detected were early malignant lesions, carcinoma *in situ* (CIS), or minimally invasive carcinoma. Many of these early esophageal cancers were not seen by the esophagoscope or radiographic examination. The early lesions were resectable and curable. A 5-year survival rate of 90.3% has been achieved in this group of patients. This experience was at great variance with the population of hospital patients in whom early cancer was exceptional. The results of cancer detection

TABLE 2-1
Accuracy of Cytodiagnosis of the Balloon Technique

	Cancer			Non-Cancer		
Reporter	Total No. of cases	Cases correctly diagnosed	Accuracy (%)	Total No. of cases	Cases correctly diagnosed	Accuracy (%)
Shen and Oiu (1963)[13]	156	137	87.8			
Zhow (1965)[21]	52	49	94.2			
Zhen and Oiu (1966)[22]	110	108	98.2			
Zhang (1969)[20]	124	114	91.9	235	234	99.6
Shu (1973)[24]	100	99	99.0	20	20	100.0
Young and Shu (1975)[25]	470	456	97.0	130	124	95.4
Zhang (1978)[23]	849	803	94.6	116	113	97.4
Total	1,861	1,766	94.9	501	491	98.0

are compared with hospital populations in Table 2-2.

INTERPRETATION OF THE TYPE AND ORIGIN OF CARCINOMA

The majority of esophageal cancers are of the squamous cell type, whereas gastric cancers are usually adenocarcinomas. The type of cancer cells often aids the physician in determining the location of the malignant lesion. One hundred ninety-eight cases of esophageal and gastric cardia cancer could be differentiated cytologically with an accuracy of 97%. The most common cause of error was poorly differentiated squamous cell carcinoma, which could be mistaken for an adenocarcinoma.

COMPARISON OF DETECTION RATES OF CYTOLOGIC, ENDOSCOPIC, AND RADIOGRAPHIC METHODS

In early esophageal cancer the performance of cytology is optimal, whereas in advanced cancer radiographic studies are optimal. Endoscopy is unnecessary in advanced cancer (see Table 2-3).

The popularity and development of cytopathology of esophageal cancer in China are attributable to the continuous improvement of collection apparatus in the last 20 years. Due to improvement in early diagnosis and surgical intervention, the 5-year survival has risen from 10 (1950s) to 30% (1970s) in advanced cancer and 90% in early cancer.

Cytology is the best method for routine detection of early esophageal cancer, followed by endoscopy. About 20% of the cases of early cancer are not seen by esophagoscopy and 40% are missed by radiography. Of course, these methods supplement each other. On the other hand, radiography is best in advanced cancer, because with a late lesion obstructing the lumen, it is impossible to pass the balloon beyond the point of obstruction; this is why no cancer cells are obtained and a false-negative reading results. Cytology, radiography, and endoscopy are used in the following manner to establish a diagnosis. If a patient has either a positive cytology or an abnormal radiograph, endoscopy is used. If both cytology and radiography are suspicious, endoscopy is performed. When both cytology and radiography are positive, endoscopy is considered unnecessary.

INVESTIGATION OF SEVERE DYSPLASIA

In the cytological mass surveys, the ratio of early esophageal cancer to severe dysplasia was 1:2 (Table 2-4). The natural history of dysplasia can be followed. In fact, 530

TABLE 2-2
Comparison of the Incidence of Early Carcinoma in Mass Survey and Hospital Patients

	Areas	Years	Total No. examined	No. of proved carcinomas	No. of early carcinomas (%)
Patients					
City hospital	Cancer Institute (C.A.M.S.)	1965–1978	7,888	1,253	3 (0.23)
	Medical College (Henan)	1963–1972	4,747	535	10 (1.90)
Countryside hospital	Linxian Hospital (Henan)	1969–1976	22,805	9,320	965 (10.35)
Mass survey of persons with symptoms	Linxian Hospital (Henan)	1963–1969	7,686	510	86 (16.86)
Mass survey of persons aged 30 or above	5th Areas of Henan Province	1970–1973	13,364	158	101 (63.92)
	Anyang Regions (Henan)	1974	23,762	171	96 (56.14)
	Yao Commune (Henan)	1974	14,002	221	171 (77.37)
	Yang Teng Commune (Shanxi)	1974–1975	5,613	71	59 (83.09)
	Sie and Mu Iom Commune (Si Chuan)	1973	5,196	27	24 (88.88)
	Shen Guan Commune	1975	19,250	232	198 (85.34)
	Total	1970–1975	81,187	880	649 (73.75)

TABLE 2-3
Comparison of Detection Rates with the Three Methods[a]

	Cytology	Endoscopy	Radiography
Early cancer	93.8–94.0%	75.0–91.7%	66.7–82.0%
Advanced cancer	87.8–99.0%	Not done	98.0–100%

[a] See text.

TABLE 2-4
Results of Mass Cytological Survey in Some Areas of Henan, Hebei, Shanxi, Sichuan (1970–1978)

No. examined	No. with severe dysplasia (%)	No. with esophageal cancer (%)
124, 829	2123 (1.70)	1190 (0.95)

cases with dysplasia have been followed for 1–12 years. The results have shown that the longer the duration of severe dysplasia, the higher the rate of malignant change.[16]

We have compared severe dysplasia with normal and mild dysplasia, with respect to malignant changes, and the degree of severity of dysplasia correlated with the malignant potential.[16]

Cytological mass surveys disclosed that the incidence of esophageal dysplasia coincided with that of esophageal carcinoma in different geographical areas. Therefore, severe dysplasia should be considered a precancerous lesion, which can be treated with good results. Thus, we think that this is a key link in the control of cancerous changes.

UNUSUAL EPIDEMIOLOGIC FEATURES

High mortality areas of esophageal cancer are located in north China. There is an irregular, concentric belt-like mountain area bordered by Henan, Hebei, and Shanxi Provinces of north China. (Mortality among the minority nationalities in China is discussed in Chapter 1.) Cytologic mass surveys disclosed that the number of early cancers found in various areas coincided with the mortality rates, i.e., high rates of discovery were present in areas with high mortality (see Table 2-5).

COORDINATION OF ETIOLOGIC RESEARCH IN ESOPHAGEAL CANCER

The etiology of esophageal cancer is still uncertain, but fungus is one factor under strong suspicion. Cytologic studies of high-risk areas revealed a high frequency of fungal infection in the esophagus that were associated with epithelial changes. Among 555 cytologic specimens,[18] fungi were found in 90% of cancer cases (90/100), 72% of the patients with severe dysplasia (180/248), and 31% with normal or atypia (64/207).

Of 155 biopsy specimens obtained during endoscopy and 30 surgical specimens, fungal infection was common. Dysplastic epithelium with fungal invasion is 10 times more frequent than normal epithelium.[19] It may be deduced that the long-term persistence of fungi with carcinogenic potential in the esophagus may promote local malignant change.

1. Esophagitis and dysplasia of the esophagus occur 10 years earlier than esophageal cancer.
2. The more severe the dysplasia, the higher the rate of severe inflammation.
3. The amount of fungi in the specimen coincided with the number of inflammatory cells.

TABLE 2-5
Result of Cytologic Mass Survey in Different Geographical Areas and Nationalities of China[a]

Areas of cytologic mass survey	Mortality of esophageal cancer	Screening survey		
		Total persons examined	Cancer	
			No. of cases	(%)
(Kazak) Manacy County Sinjiang	240.70	579	11	1.90
Yoatsun Commune Linxian, Henan	169.28	14,002	221	1.58
Yang Cheng Shanxi	161.90	3,316	48	1.44
Linxian, Henan	111.00	7,212	78	1.07
Yang Tang Szechuan	79.00	5,697	52	0.91
Zhoung Shang, Hubei (Migrants from Henan)	74.00	13,613	60	0.44
Anyang, Henan	71.12	17,471	88	0.50
Min Chong Szechuan	68.34	5,169	27	0.52
Shanghai	10.67	878	0	0
Zhoung Shang, Hubei (non-migrants)	7.68	6,909	3	0.04

[a] Age-adjustment mortality of China per 100,000 people.

CHAPTER 3

Diagnostic Techniques and Criteria

THE APPARATUS FOR SECURING CYTOLOGIC SAMPLES FROM THE ESOPHAGUS

In 1950, Panico, Papanicolaou and associates[26] proposed using a balloon covered with silk mesh to improve the cytologic detection rate of gastric cancer. A similar but improved apparatus was used by Brainsma in 1957 to secure cytologic samples from the esophagus.[27] An improved version of Brainsma's balloon, provided with a double-barrelled collector, was devised by Shen and associates[13] in 1959 and has been in use since that time. This collector consists of a balloon covered by a mesh made of surgical silk thread with a rubber tube. Within the rubber tubing there is a main tube and an auxiliary tube. The main double-barrelled tube is 60 cm in length and 0.3 cm in diameter. One barrel is used for the passage of air and the other for suction of fluid in the esophagus or stomach. The ovoid balloon is made of thin rubber and has a length of 5 cm and the largest diameter of 2 cm. About 20 ml of air can be instilled into the balloon (Fig. 3-1*B*).

In 1965, further improvements in the instrument were made at the Cancer Institute Chinese Academy of Medical Sciences in Beijing by attaching a single-barrelled plastic tubing to the mesh covered balloon (Fig. 3-1*A*: Type 72, left of upper and middle row; Type 65, middle of middle and lower row).[25]

Shu and associates[14] developed a smaller balloon that was easily swallowed and more readily accepted than the previous technique, thus improving further the collecting technique. Recently, a still smaller balloon has been introduced for the comfort of the patient. Type 72, with a luminal diameter of 2.5 mm and a balloon diameter of 2.5 cm, has been replaced by Type 75, with diameters of 1.5 mm and 0.8 cm, (Figs. 3-1*A*: middle of upper row and Fig. 3-1*D*) and Type 77, with diameters of 1.5 mm and 0.4 cm, (Figs. 3-1*A*: right of upper row and Fig. 3-1*E*).

Shu and associates[24] of Fujian Medical College, using an esophageal balloon of their own design, rinsed the collected specimen in normal saline, and after centrifugation prepared cytologic smears on 120 cases (100 cancers, 20 noncancers), and missed diagnosing only one case of cancer.

The Cancer Detection and Prevention Centre in Yanting County, Sichuan, and related institutes,[28] devised a "Type 76 cytologic collector for esophagus and cardia." The outer silk mesh was eliminated and the balloon was manufactured with a coarse, granular outer surface. The balloon was small, measuring only 2 cm in length with a maximal diameter of 1 cm. It had two small compartments, one at the top of the balloon and

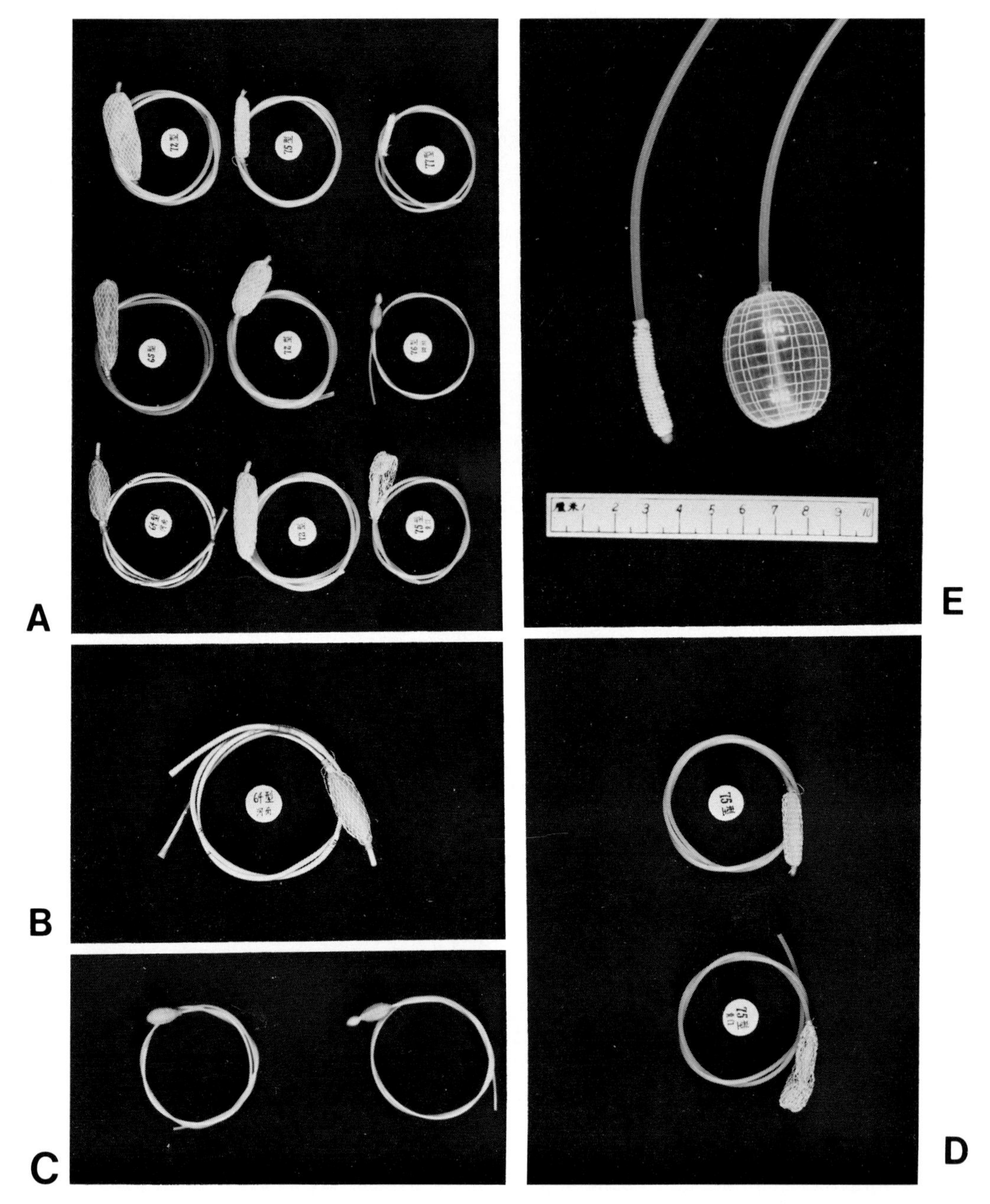

Fig. 3-1. Apparatus. (*A*) Various types of balloons for collection of esophageal specimens in China. (*B*) A double-lumen rubber tube with an inflatable abrasive balloon covered by a mesh net. (*C*) A single-lumen plastic tube with an inflatable balloon, but without the mesh net (Type No. 76). (*D*) A single lumen plastic tube with an inflatable balloon covered by a mesh net (Type No. 75). (*E*) A single lumen plastic tube with an inflatable balloon covered by a mesh net (Type No. 77).

the other within the balloon itself. These compartments could be filled with barium for localization under x-ray, and when air was injected, the balloon would swell without disturbing the barium in the compartments. More than 5000 persons have been examined with this apparatus, with results comparable to those obtained with the silk mesh balloon. The advantage of the new collector is that it is less painful to insert and, therefore, more readily accepted by patients (Fig. 3-*A*, Type 77, upper right, and Fig. 3-1*C*).

CLINICAL METHODS

Routine Examination of the Esophagus by Balloon

Precautions: Before the examination by balloon is undertaken, the medical history, the results of physical examination, and laboratory data must be obtained. Contraindications to the esophageal sampling are: liver cirrhosis with esophageal varices, gastric ulcer with active bleeding, severe heart disease or hypertension, acute laryngitis, overly weak constitution, and a barium swallow within 2 days prior to examination.

Procedure

1. The procedure and its purpose must be carefully explained to the patient.
2. The patient should fast overnight.
3. Prior to examination, all dental prostheses must be removed and a vigorous oral rinse performed.
4. The patient is placed in a comfortable sitting position with the head in an upright manner, neither flexed nor extended.
5. The patient is asked to say "aah". The balloon is then inserted into the back of the throat and the patient is asked to swallow. The balloon (with the attached tube) is rapidly passed down the esophagus until it enters the gastric cardia. This sequence is important for successful passage.
6. The balloon is inflated (20 ml of air for type 75 instruments, 30 ml for type 77). If resistance is encountered, the balloon is slowly withdrawn with some release of air. Air should be reintroduced, however, after passage through the narrow channels. When the balloon is 18 cm from the front teeth, all air should be released and the balloon quickly removed.
7. Smears are made and fixed in 95% ethanol, stained by Papanicolaou's method, and microscopically examined.
8. Shu[24] also rinsed the balloon in normal saline. From this centrifuged specimen, cytologic smears were prepared with only 1% of cancers missed.

Localization of Esophageal Lesions

Routine barium swallow is often unable to diagnose early esophageal cancers. Preoperative localization of the lesion is of paramount importance, as exploration of the entire organ at surgery is time-consuming and predisposing to intercurrent infections. Consequently, localization of small or early lesions by cytology prior to surgery is of great practical significance. We have adopted the segmental cytologic localization method. Proceeding downward from the 20 cm mark, the examination proceeds in steps of 5 cm. At each level a sampling is performed until the depth of 35–38 cm has been reached. The point of transition, from the negative to positive cytologic sample, is taken as the upper border of the lesion (Fig. 3-2). The correlation between cytologic localization prior to surgery and the pathologic findings in the resected specimen has been studied. Of the 129 patients with early esophageal cancer who underwent surgery in the People's County Hospital at Linxian, the accuracy rate of cytology sampling was 96.9% (Table 3-1). This method of cytologic localization was used when the lesion could not be localized by endoscopy or x-ray, as summarized in the scheme.

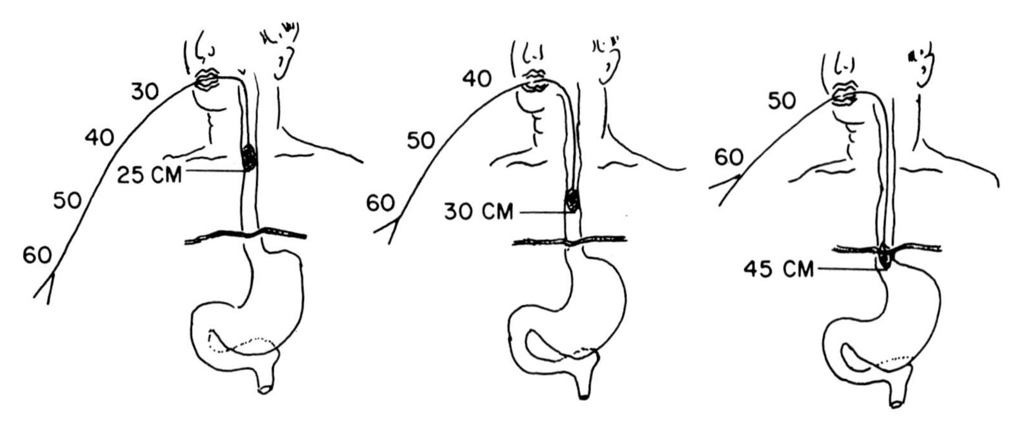

Fig. 3-2. Localization of the lesion by balloon technique. (From Koss and Coleman: *Advances in Clinical Cytology, Vol. 2,* Masson Publishing USA, Inc., New York, 1984; reproduced with permission.)

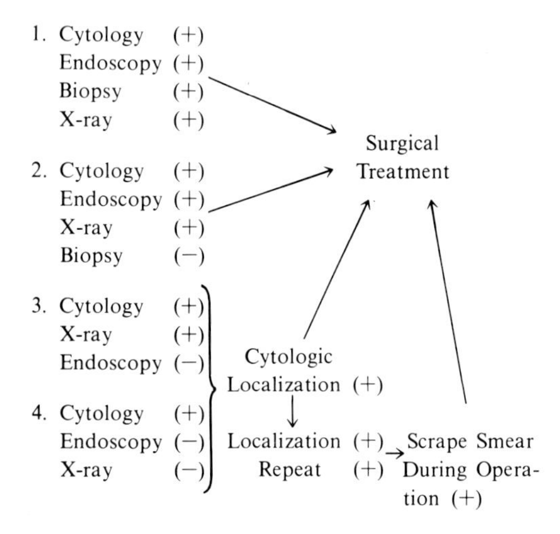

DIAGNOSTIC APPLICATIONS OF MORPHOLOGIC CRITERIA

Grading of Cytological Findings

According to the morphological characteristics, particularly the configuration of the nucleus, the squamous epithelial cells of the esophagus may be divided into five grades (Figs. 3-4, 3-5, 3-6, 3-7).[15]

Normal: The majority are intermediate cells, and about 10% are superficial cells. Parabasal cells are very rare (Figs. 3-3*D* and 3-3*Ea, b*).

Atypia: The majority of the cells are intermediate and superficial cells. The nuclear enlargement is slight, the chromatin is finely granular, and the shape of the nucleus is round or oval. Sometimes there is a greater number of superficial cells and basal cells. Inflammatory cells are also frequently seen (Figs. 3-8 and 3-9).

Dysplasia: Dysplastic cells are divided into two classes:

1. *Mild dysplasia*—Nuclear hyperchromatism occurs in cells derived from the intermediate and superficial cells. The

TABLE 3-1
Accuracy Rate in Location of Cytology and Radiographs[a]

Gross specimens (histologically confirmed)	Cytology		Radiograph	
	No. of cases	(%)	No. of cases	(%)
Accurate localization	125	(96.9)	79	(61.2)
Inaccurate localization	4	(3.1)	50	(38.8)
Total	129	(100.0)	129	(100.0)

[a] Linxian Hospital, Henan.

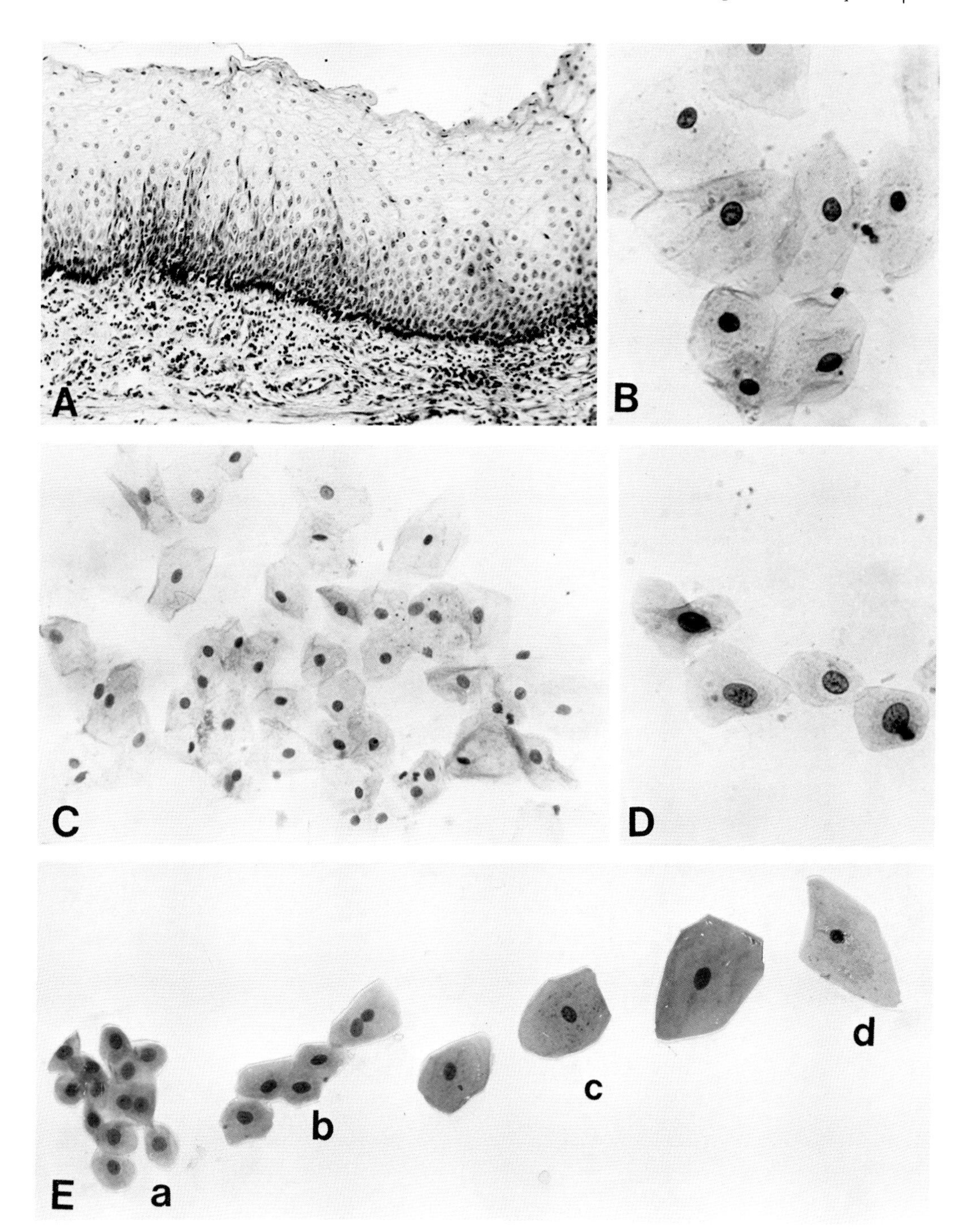

Fig. 3-3. Normal epithelium of the esophagus (*A*) histologic pattern (×90); (*B*) intermediate cells (×180); (*C*) intermediate cells and superficial cells (×90); (*D*) parabasal cells (×180); (*E*) parabasal cells (*a,b*); intermediate cells (*c*); superficial cells (*d*) (×180).

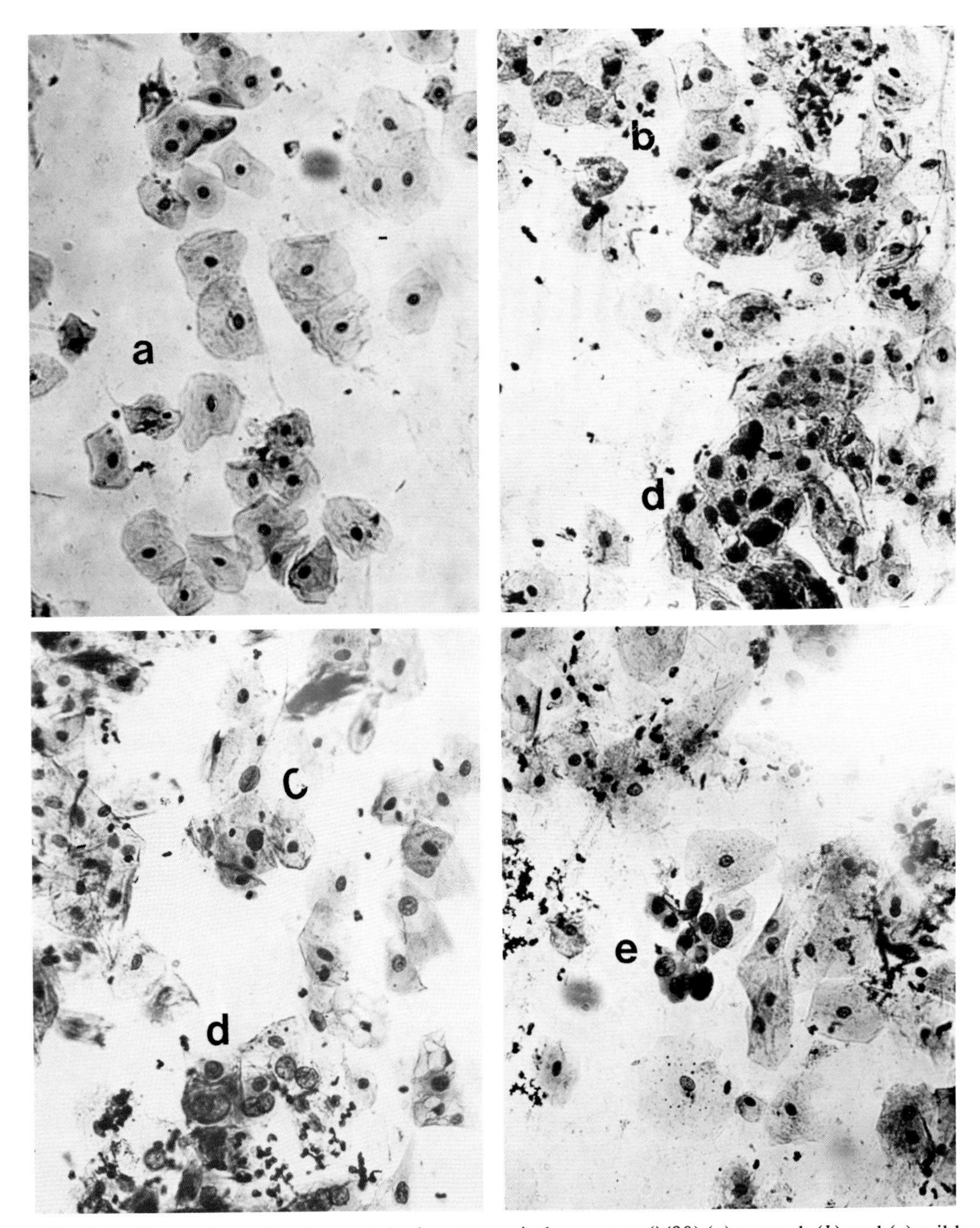

Fig. 3-4. Comparison of various cytologic patterns in low power (×90) (*a*) normal; (*b*) and (*c*) mild dysplasia; (*d*) severe dysplasia; (*e*) cancer.

nucleus is enlarged but not more than three times that of the normal intermediate nuclei (Figs. 3-10 and 3-11).

2. *Severe dysplasia* (*Grade I*)—The size of the nucleus of the dysplastic cell is greater than three times that of the normal intermediate cell nucleus. Hyperchromatism is more marked. Dysplastic parabasal cells are increased in number (Figs. 3-12 and 3-13*A*,*B*).

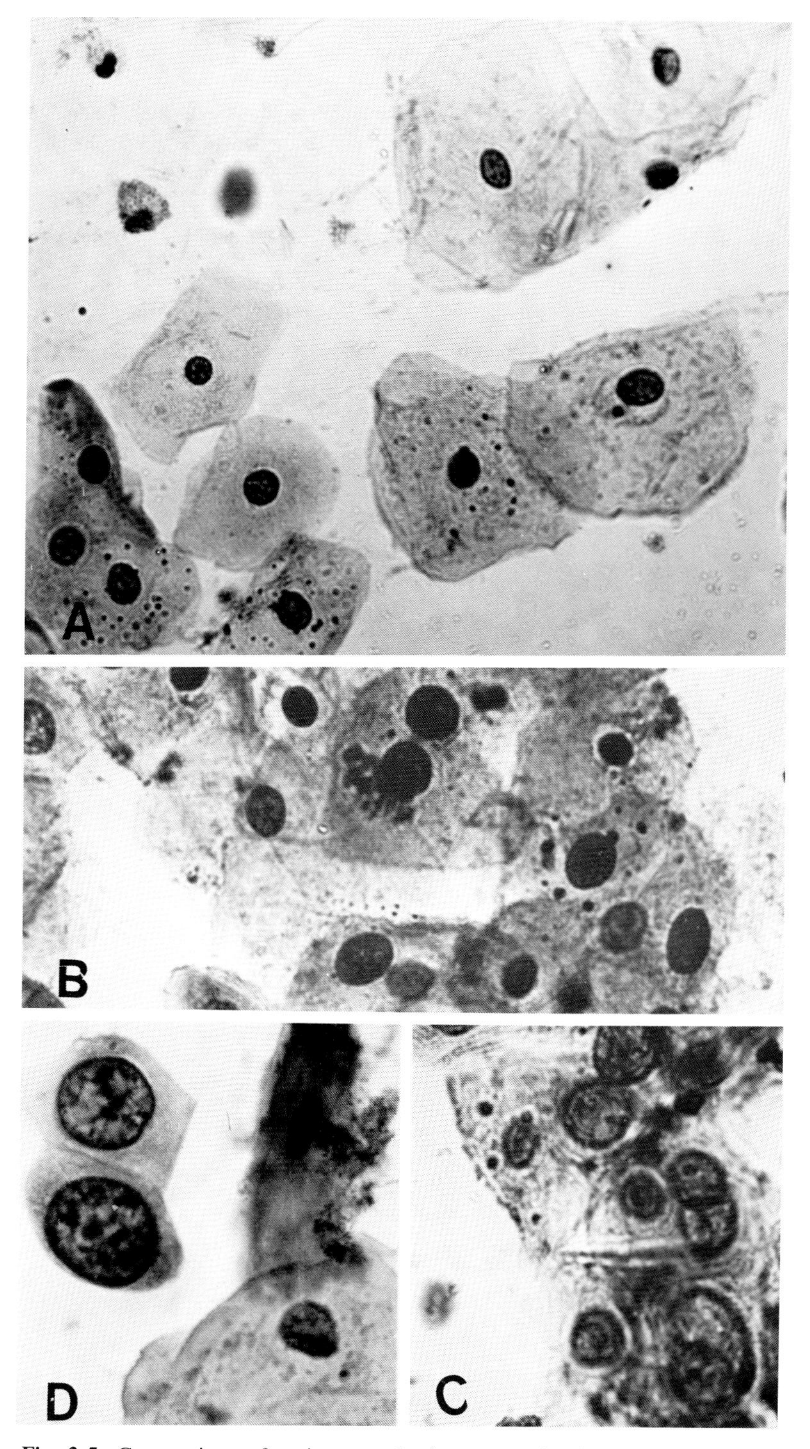

Fig. 3-5. Comparison of various cytologic patterns in high power (×400) (*A*) normal (top half) and mild dysplasia (lower half); (*B*) severe dysplasia (Grade I); (*C*) severe dysplasia (Grade II); (*D*) cancer.

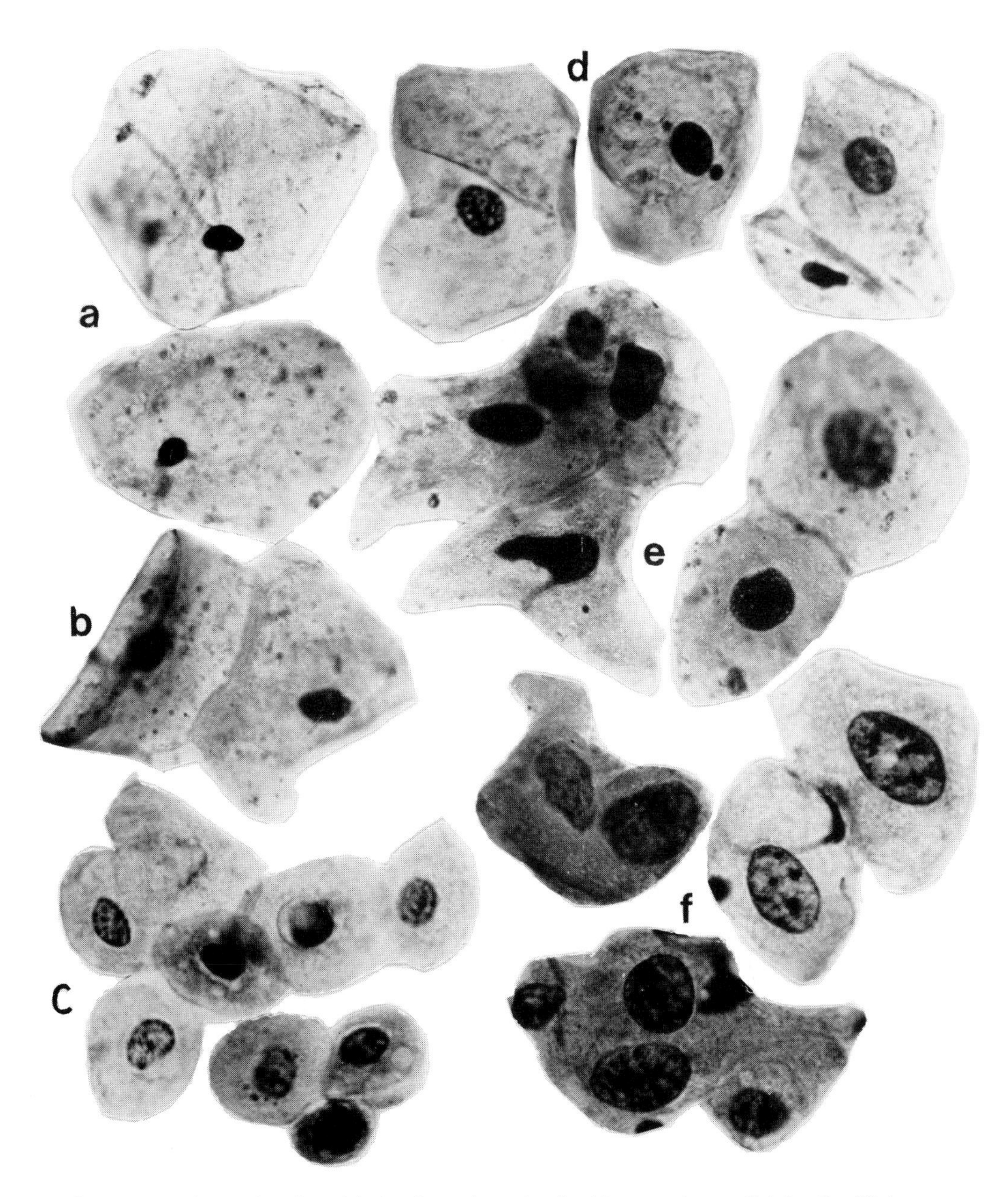

Fig. 3-6. Comparison of various kinds of esophageal cells (*a*) normal superficial cells; (*b*) intermediate cells; (*c*) parabasal cells; (*d*) mild dysplastic cells; (*e*) severe dysplastic cells; (*f*) suspicious cancer cell.

Suspicious for Cancer (*or Severe Dysplasia, Grade II*): The cytological pattern is similar to severe dysplasia, but with a greater number of dysplastic cells. Large numbers of dysplastic parabasal cells and pleomorphic dysplastic cells are present, i.e., intermediate cells with giant nuclei, fiber or spindle forms with marked nuclear hyperchromatism, and parabasal cells with nuclear hyperchromatism and irregularity. The various dysplastic

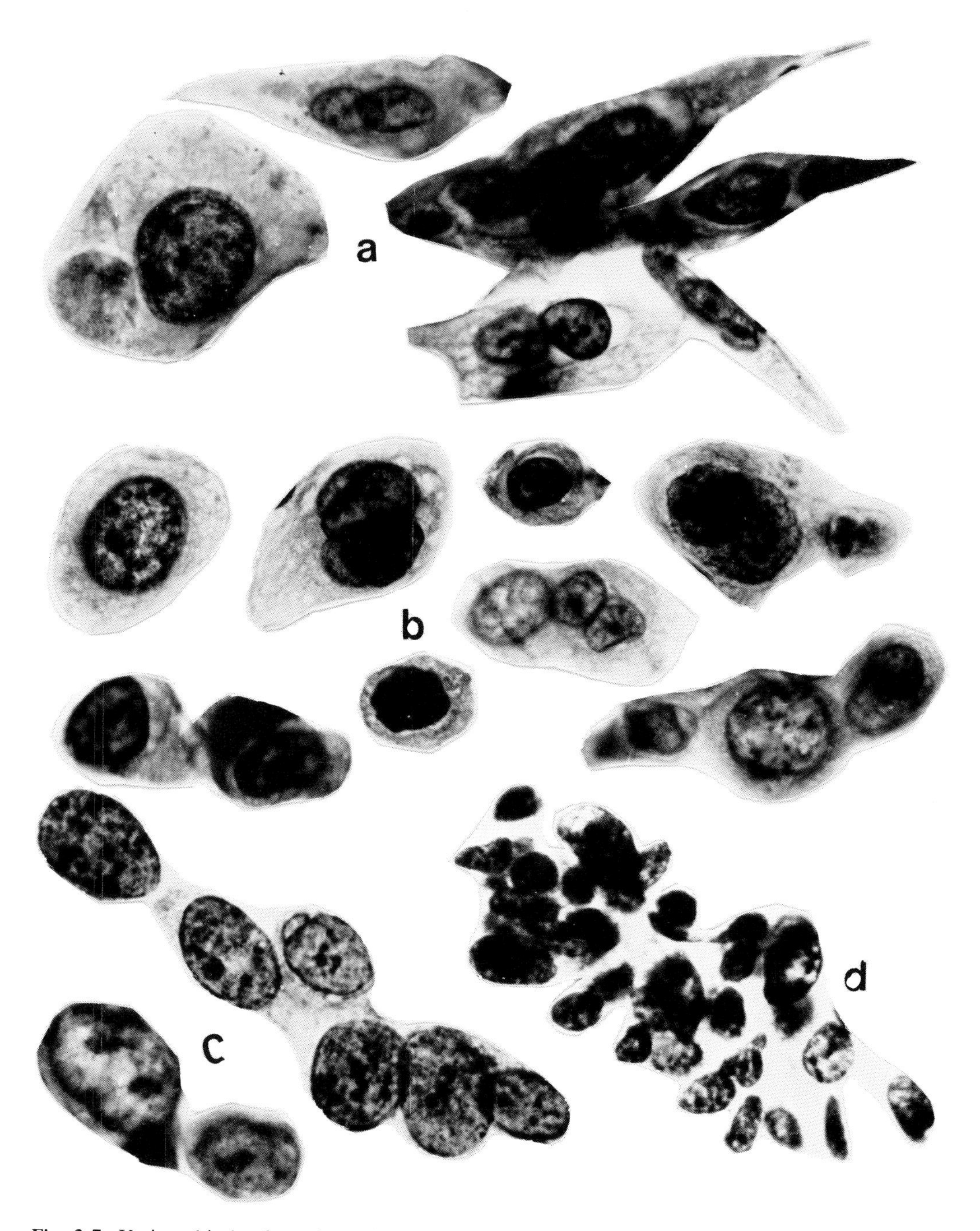

Fig. 3-7. Various kinds of esophageal cancer cells (*a*) well-differentiated cancer cell; (*b*) moderately differentiated cancer cells; (*c*) poorly differentiated cancer cells (large cells); (*d*) poorly differentiated cancer cells (small cells).

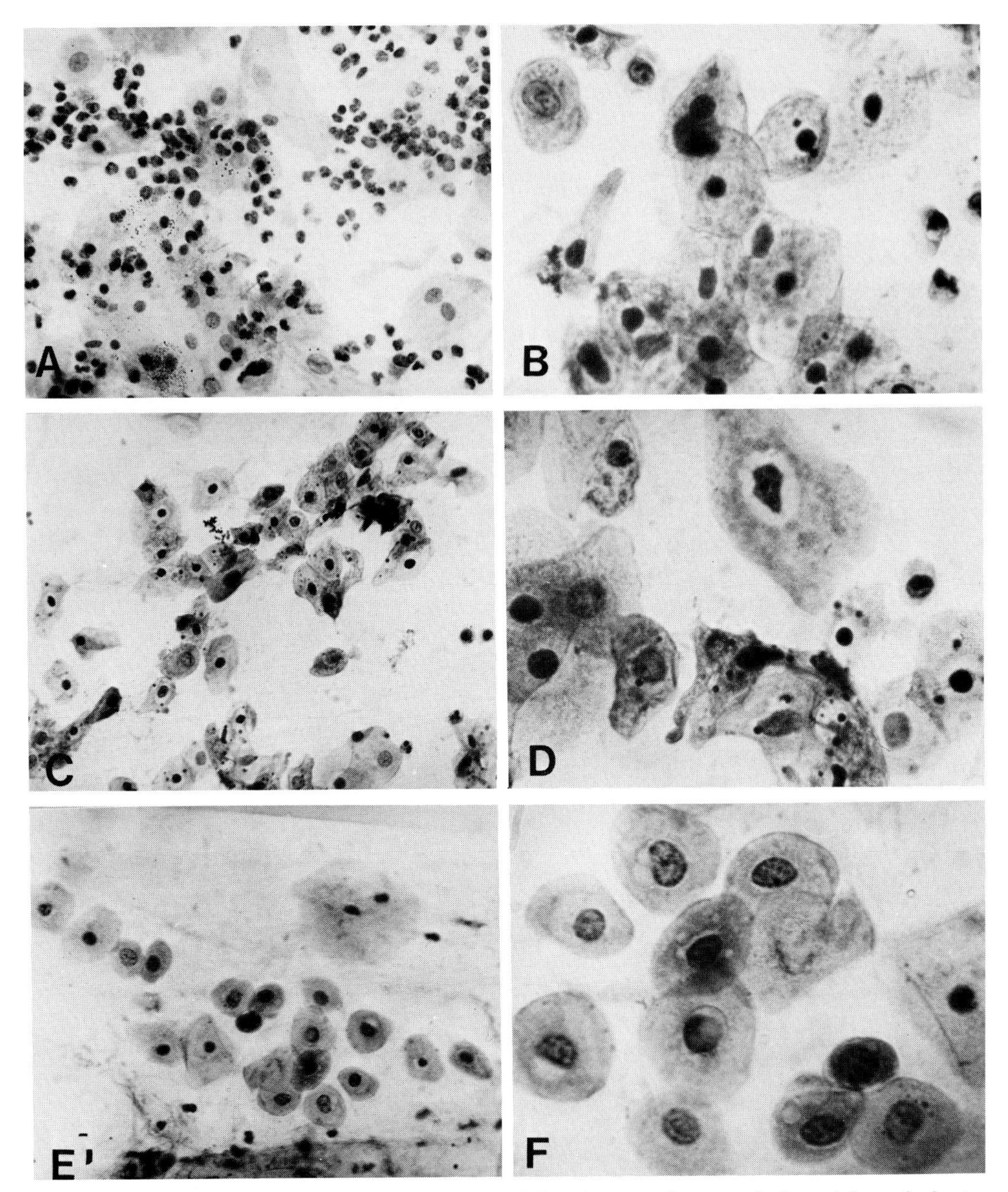

Fig. 3-8. Inflammatory cytologic patterns (*A*) cellular degeneration: cytolysis and karyolysis, inflammatory cell increase (×170); (*B*) cellular degenerative changes. Cytoplasmic swelling with opaque nuclei and irregular shapes (×340); (*C*) karyopyknosis (×85); (*D*) cellular deformity with perinuclear halo (×340); (*E,F*) increase in parabasal cells (E: ×85; F: ×340).

cells might suggest an underlying squamous cell carcinoma (Figs. 3-13*C* to *F*, 3-14, and 3-15).

Cancer: There are squamous carcinoma cells scattered through a background of dysplastic cells. The cancer cells are usually single, polygonal, and comparatively uniform in shape with malignant nuclear characteristics (Figs. 3-16, 3-17, and 3-18). Squamous carcinoma cells are divided into two types:

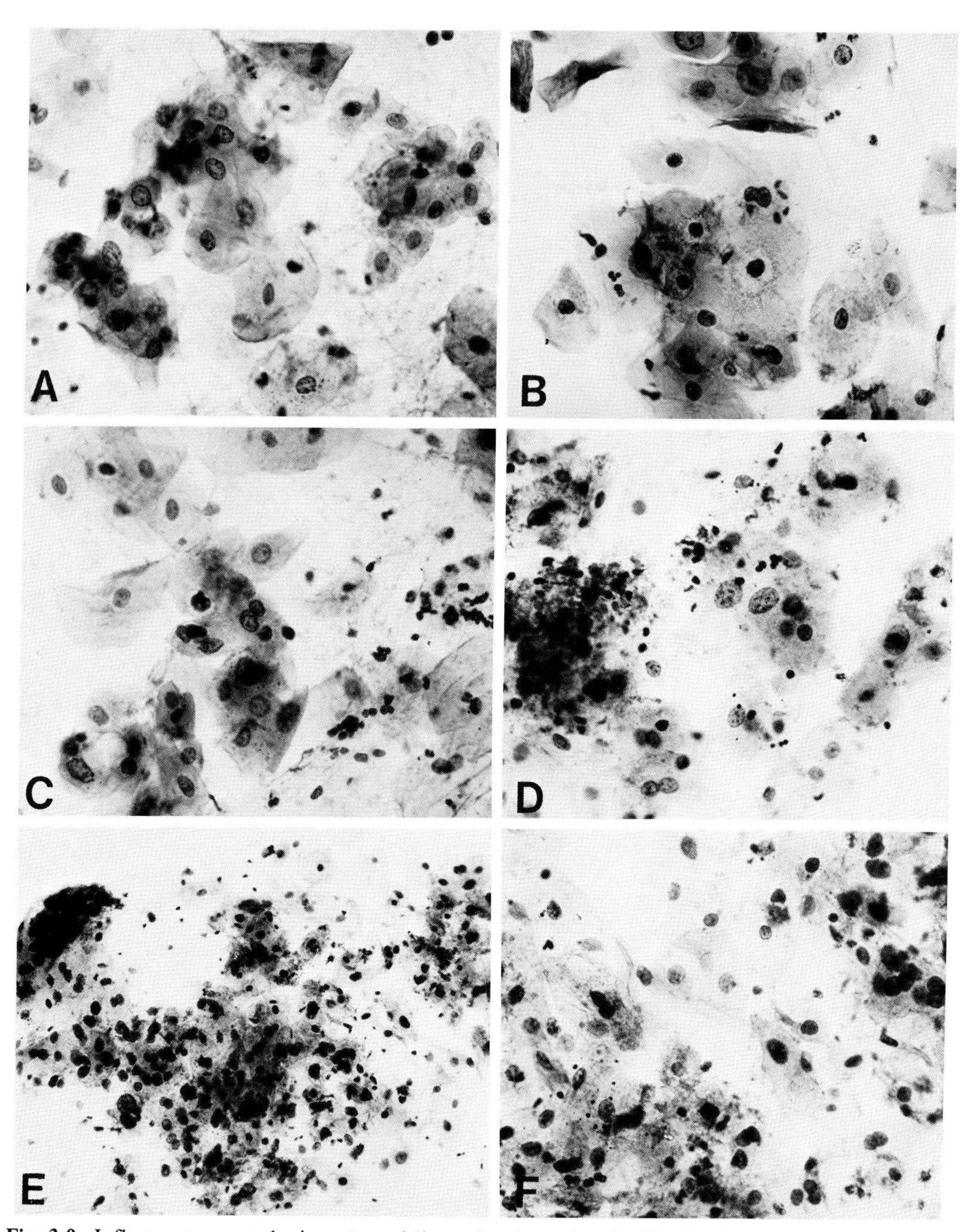

Fig. 3-9. Inflammatory cytologic pattern (*A*) atypia of parabasal cells. Nuclear enlargement and hyperchromasia (×180); (*B*) upper: nuclear hyperchromasia (×180); (*C*) lower: cytoplasmic degeneration (×180); (*D*) cytolysis and nuclear swelling (×180); (*E,F*) cytolysis and nuclear hyperchromasia (*D:* ×90; *F:* ×180).

1. *Early cancer*—Cancer cells are few in number, usually round or oval in shape, and always single and scattered. Well-differentiated cancer cells are predominant and are accompanied by large numbers of dysplastic cells. The background of the slide is "clean."
2. *Advanced cancer*—Cancer cells are nu-

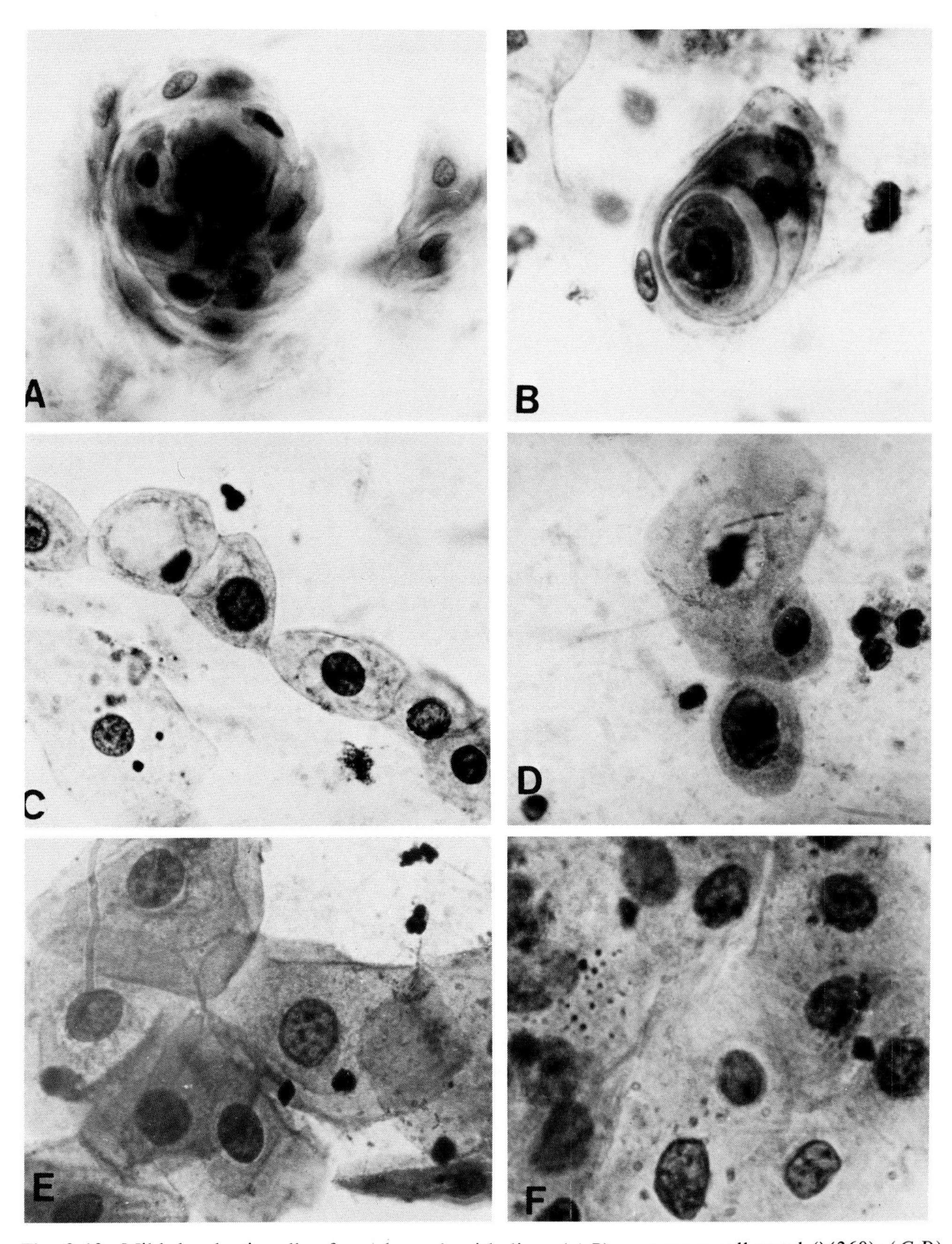

Fig. 3-10. Mild dysplastic cells of esophageal epithelium (*A,B*) squamous cell pearl (×360); (*C,D*) mild dysplasia of parabasal cells (×360); (*E,F*) mild dysplasia of intermediate cells (×360).

merous, often appearing in clusters. Pleomorphic cancer cells are abundant, particularly in keratinizing squamous cell carcinomas. Dysplastic cells are few in number and the background is "dirty."

Cytologic Classification of Squamous Cell Carcinoma of the Esophagus: Cytologic grading can help in planning treatment depending on the cell type, the degree of cellular differentiation (amount of cytoplasm), the border of the cell, and the nucleocyto-

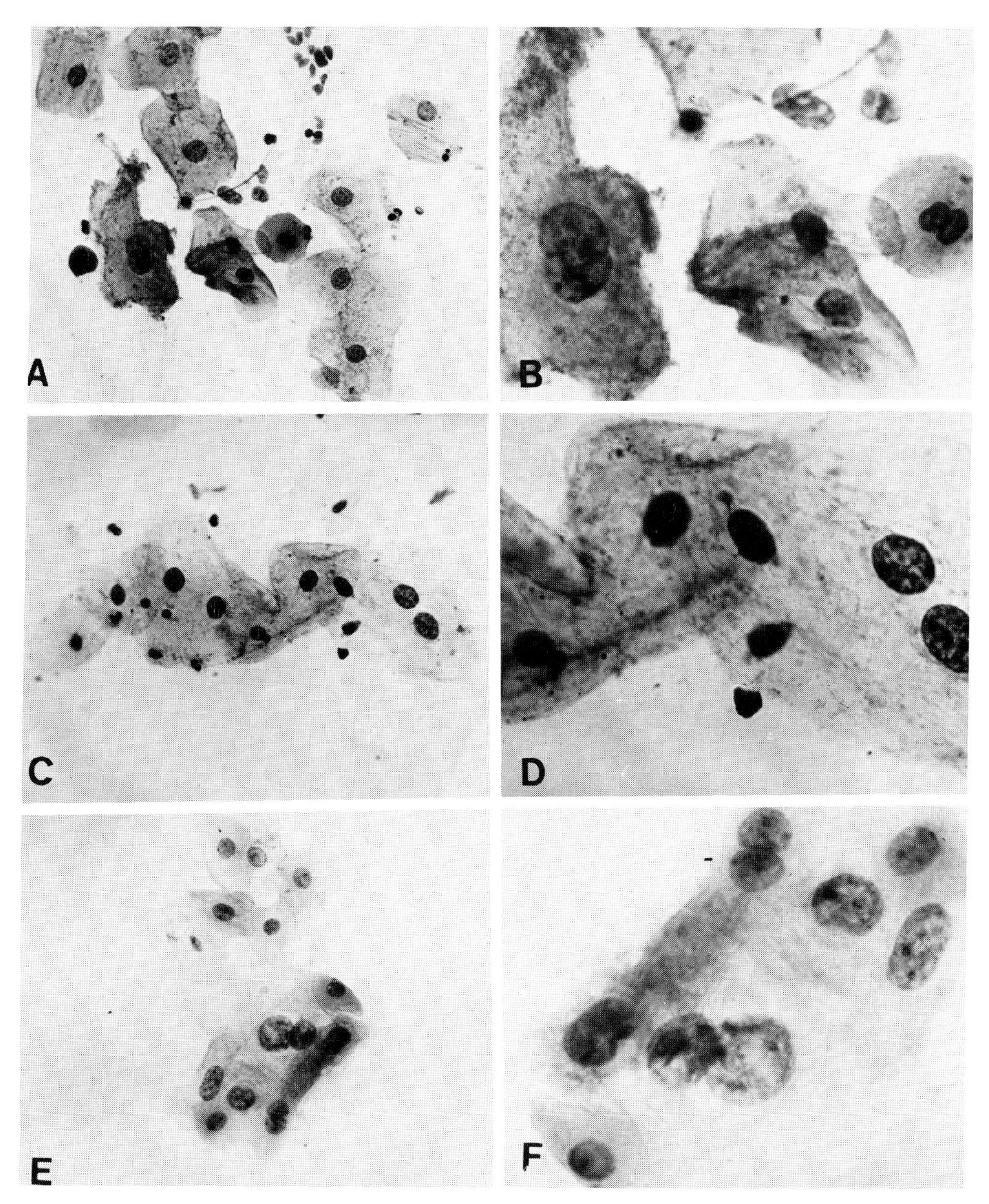

Fig. 3-11. Mild dysplastic cells of esophageal epithelium (*A,B*) mild dysplasia of intermediate cells except binucleated giant cells (*A:* ×85; *B:* ×340); (*C–F*) mild dysplasia of intermediate cells (*C,E:* ×85; *D,F:* ×340).

plasmic ratio. The squamous carcinoma cells are divided into three cytologic types:

1. *Well-differentiated squamous cell carcinoma*—The cells are mostly polygonal or bizarre-shaped with abundant cytoplasm, which stains red or orange. The nuclei are large and contain deep-staining chromatin. The nuclear structures can be either opaque (India ink) or clear. The nucleocytoplasmic ratio is slightly increased. Spindle or fiber cell

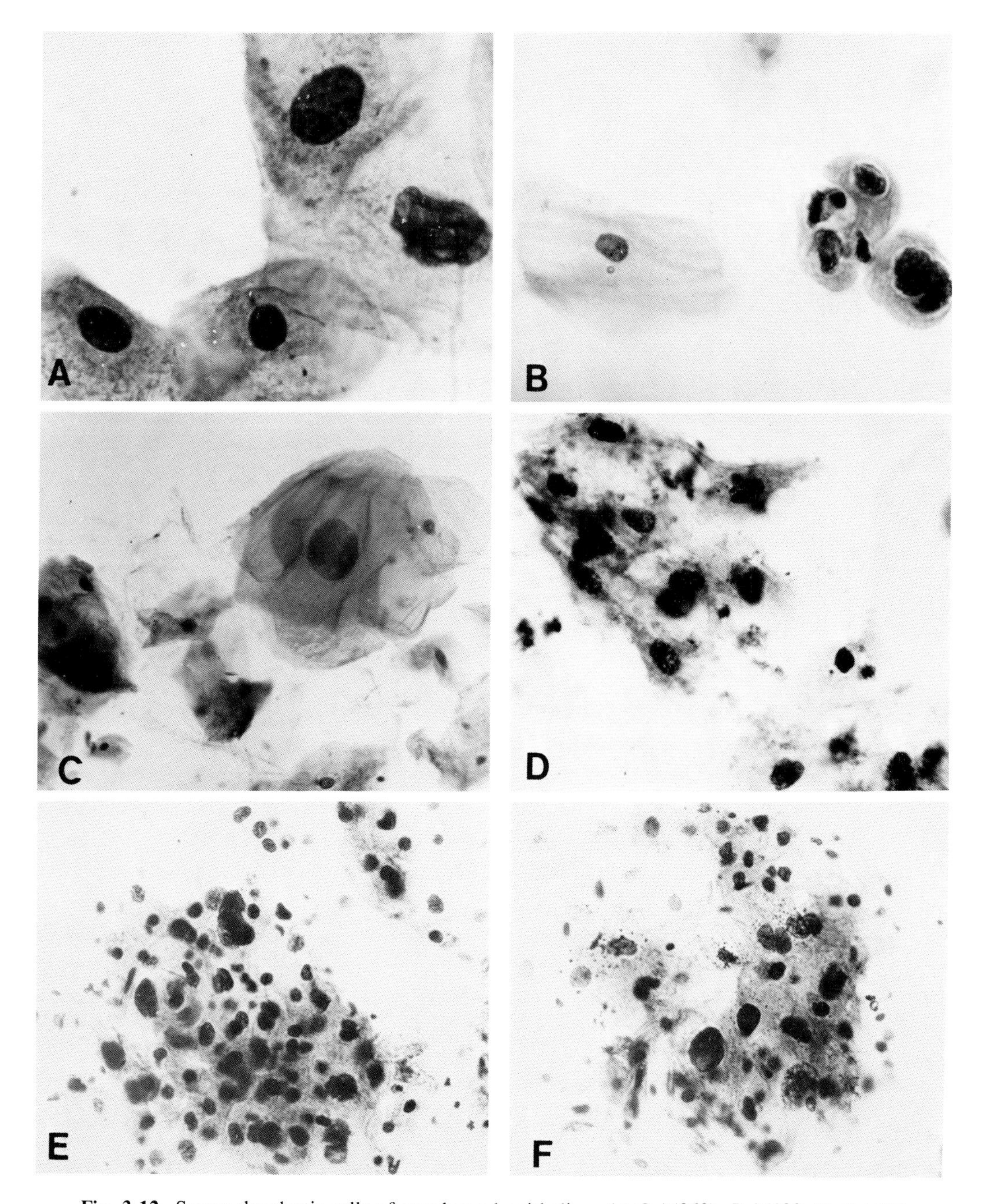

Fig. 3-12. Severe dysplastic cells of esophageal epithelium (*A,C:* ×360); *B:* ×180; *D–F:* ×90).

types are present, as well as polygonal or tadpole shapes (Figs. 3-16 and 3-7*A*).

2. *Moderately differentiated squamous cell carcinoma*—These cells are mostly round to oval and resemble parabasal cells. The cytoplasm is relatively abundant and stains a bluish color. The nuclear structures are clear and the nucleocytoplasmic ratio is moderately increased. One or several nucleoli may be observed. These cells are most common

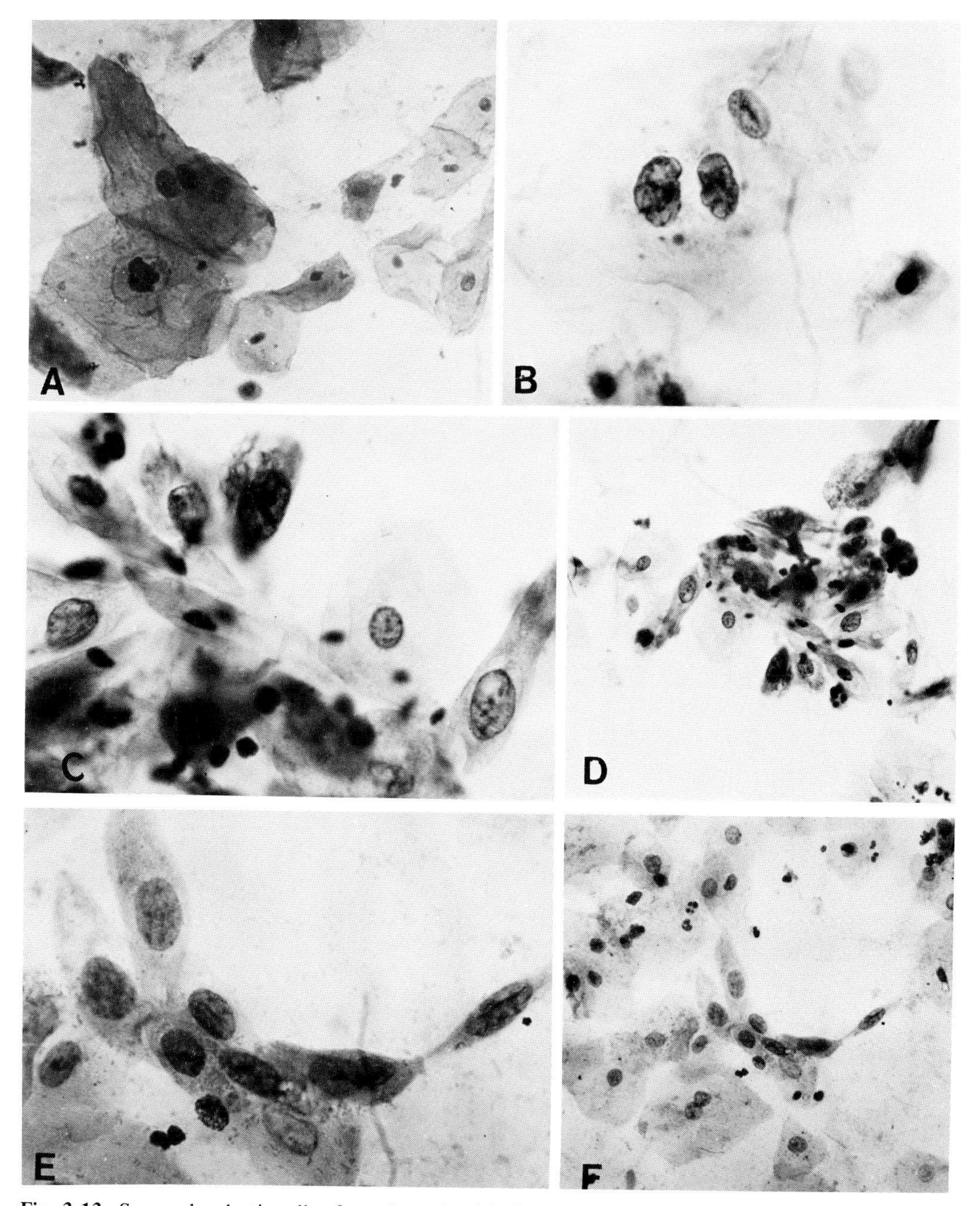

Fig. 3-13. Severe dysplastic cells of esophageal epithelium (*A,B*) large superficial cells of severe dysplasia (*A:* ×90; *B:* ×360); (*C–F*) severe dysplastic cells (Grade II) spindle shapes (*C,E:* ×360; *D,F:* ×90).

in the smear of early cancer (Figs. 3-17*A*,*B* and 3-7*B*).

3. *Poorly differentiated squamous cell carcinoma*—These cells are usually small and show great variation in size. There are two distinct presentations:

a. The cells are round or oval in shape and the cytoplasm is scanty. The

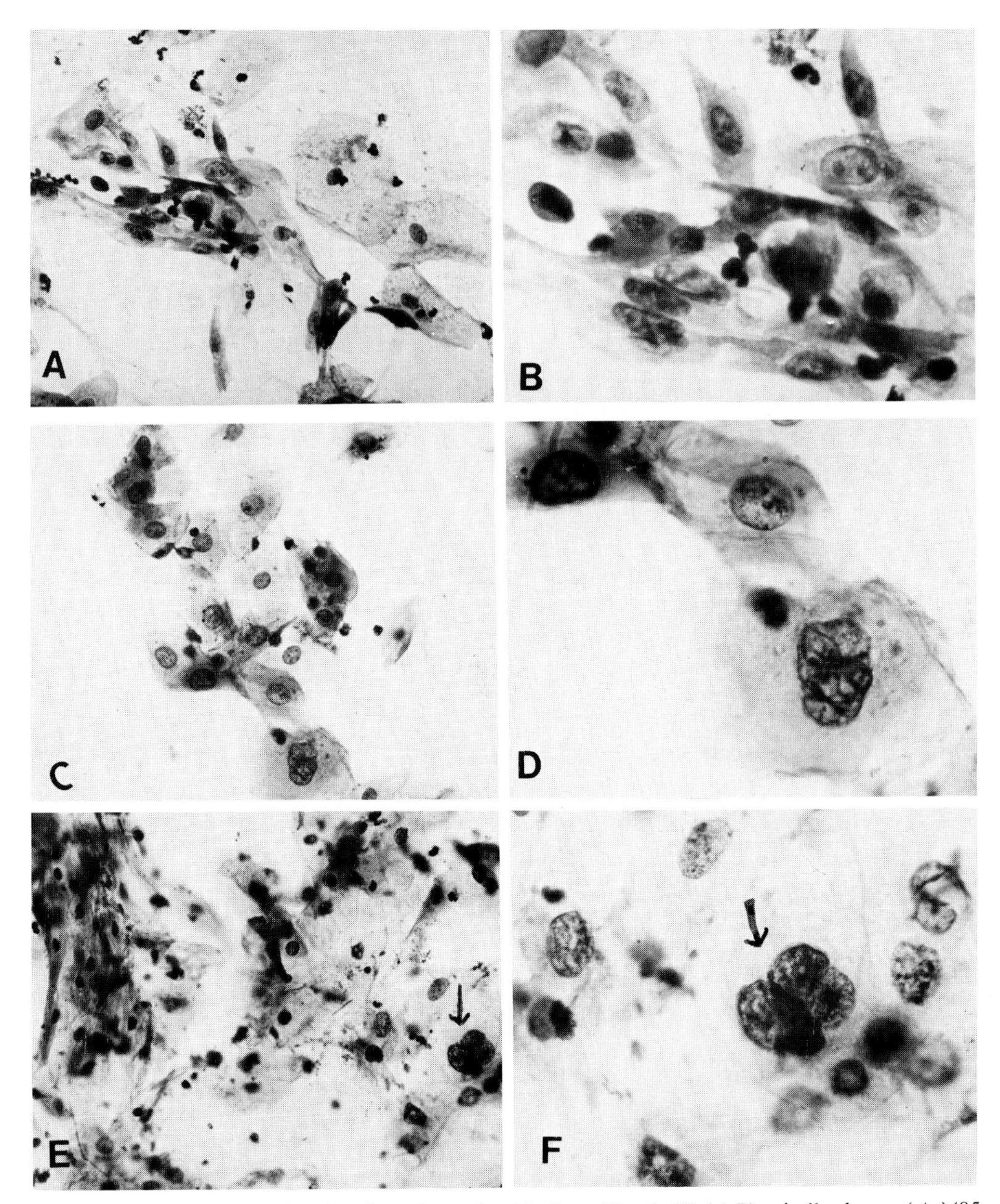

Fig. 3-14. Severe dysplastic cells of esophageal epithelium (Grade II) (*A,B*) spindle shapes (*A:* ×85; *B:* ×340); (*C,D*) severe dysplastic cells (*C,* ×85; *D:* ×340); (*E,F*) multinucleated cells (*E:* ×85; *D:* ×340).

cellular border and the nuclear border are well-defined, and the nucleocytoplasmic ratio is greatly increased (Fig. 3-17*D*). Nucleoli can be seen (Fig. 3-17*C*).

b. A more common pattern is characterized by an irregular shape. The cellular border is not well defined and naked nuclei may be observed. The chromatin is clumped and stains very deeply. Nucleoli are usually absent. Two types are seen: the rare small cell type (Figs. 3-7*D* and 3-18*A* to *C*) and the large cell type (Figs. 3-7*C* and 3-18*D,E*).

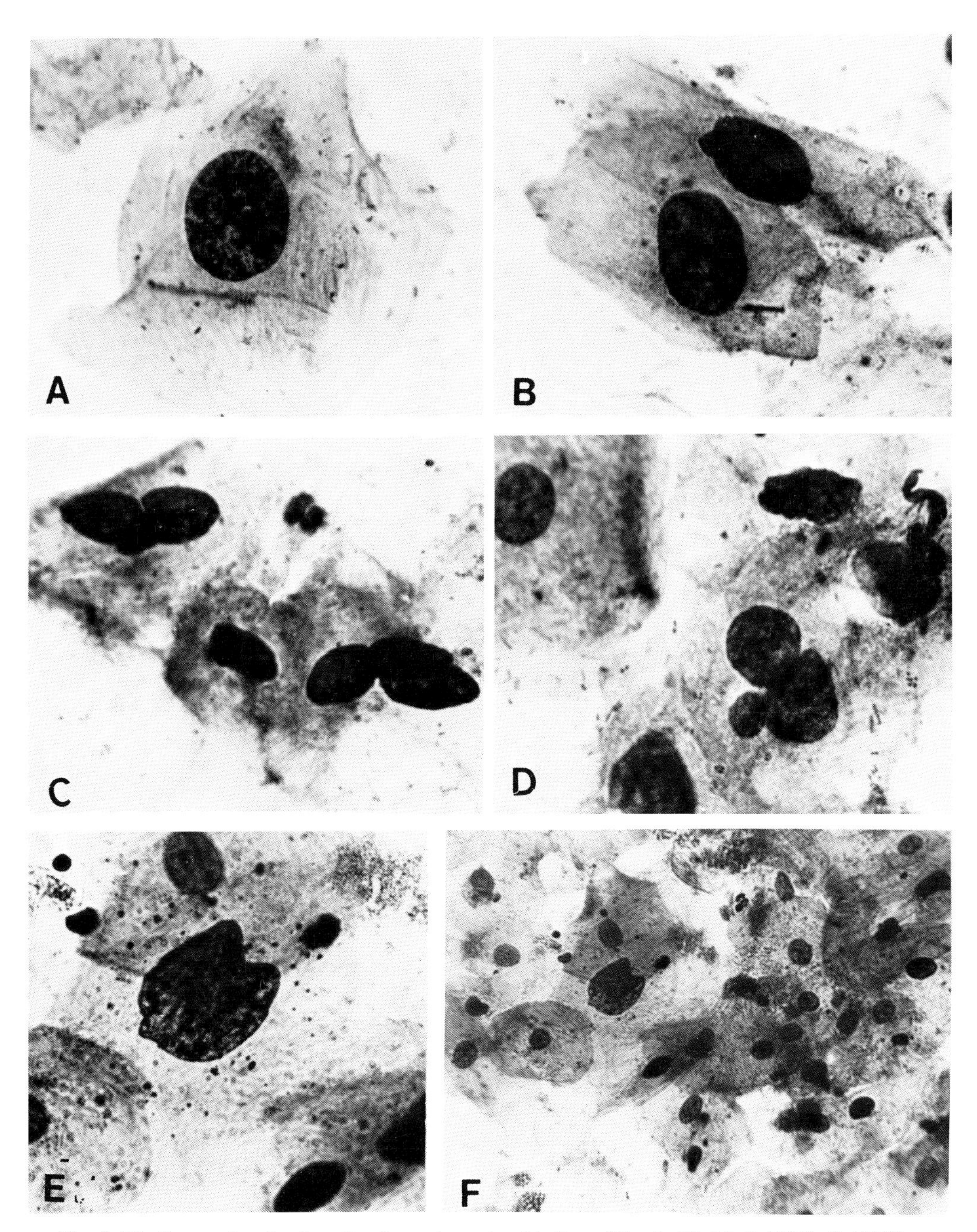

Fig. 3–15. Severe dysplastic cells of esophageal epithelium (Grade II) (*A–E:* ×540; *F:* ×180).

Grading of Histological Findings

Normal Epithelium: The esophagus is lined by stratified squamous epithelium; from the basement membrane to the lumen it consists of basal, parabasal, intermediate, and superficial cell layers. The lamina propria, consisting of fibrovascular tissue, extends as "papillae" into the epithelium.

The thickness of the epithelium is between 150 and 250 μm, and there are 15–25 cell

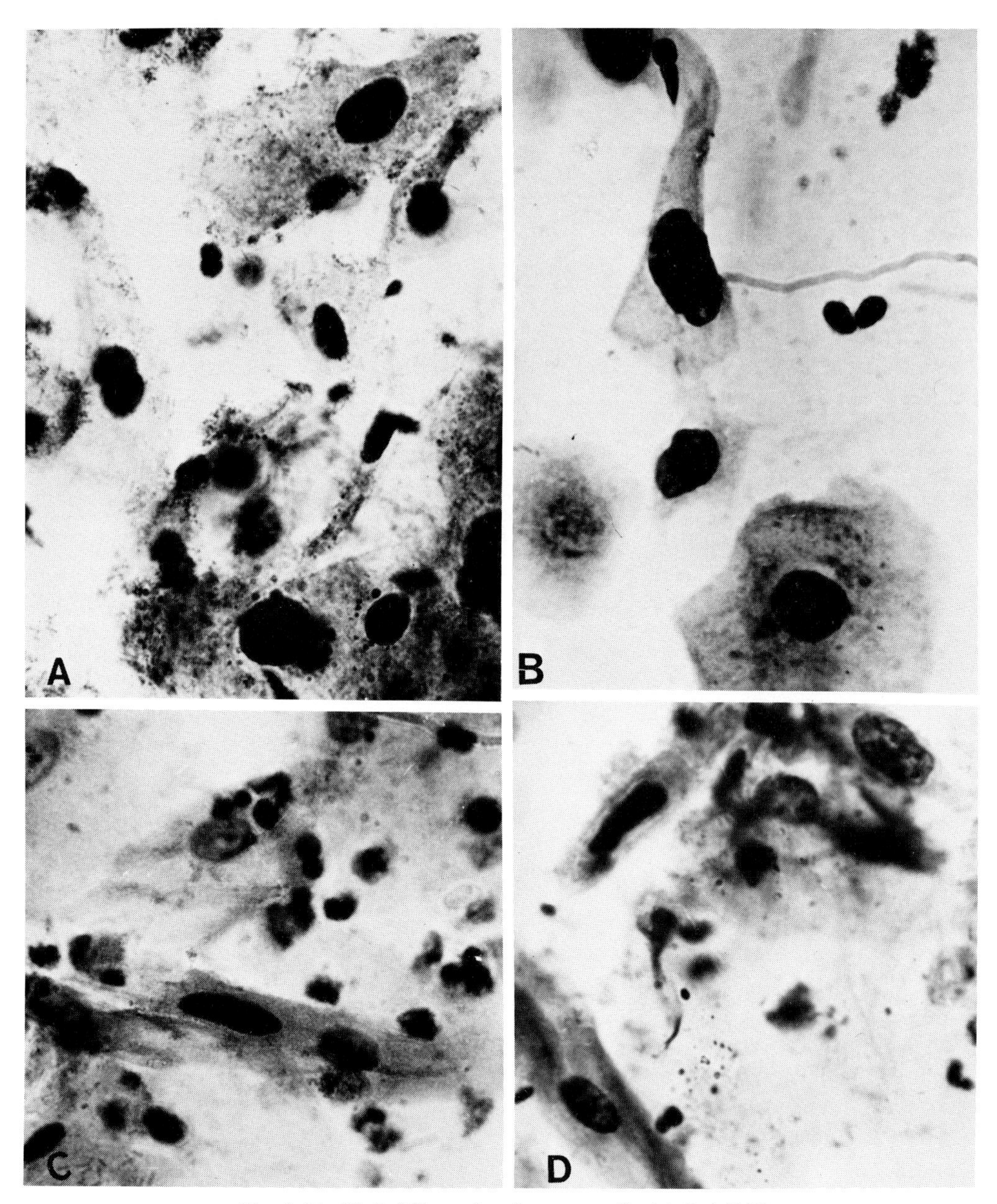

Fig. 3-16. Well-differentiated cancer cells (*A–D:* ×340).

layers in all. The basal layer is in contact with the basement membrane, and is composed of a single row of small round or oval cells. The parabasal layer consists of 2–4 layers of slightly larger round or oval cells. There are 10–15 layers of uniform, polygonal intermediate cells, which are rich in glycogen. The superficial cells consist of 3–5 layers of flattened cells with pyknotic nuclei, which represent the final product of cellular maturation. Full keratinization does not occur in the normal esophagus (Figs. 3-3*A* and 3-19*A*).

Simple Hyperplasia: The squamous epithelium becomes thicker. The increase in the cell layers is variable, cell polarity is maintained, and normal cell maturation is evi-

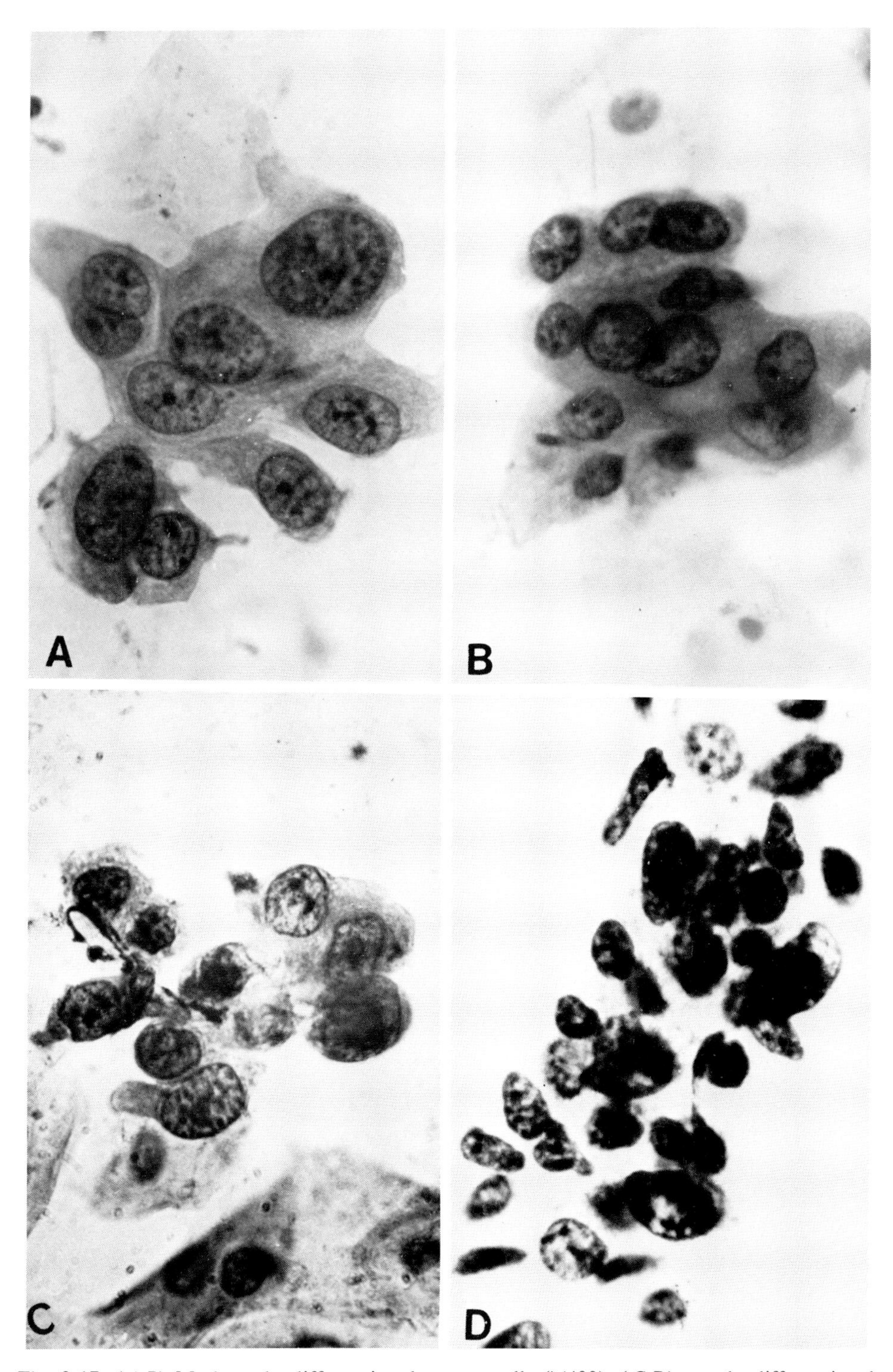

Fig. 3-17. (*A,B*) Moderately differentiated cancer cells (×400); (*C,D*) poorly differentiated cancer cells (×400).

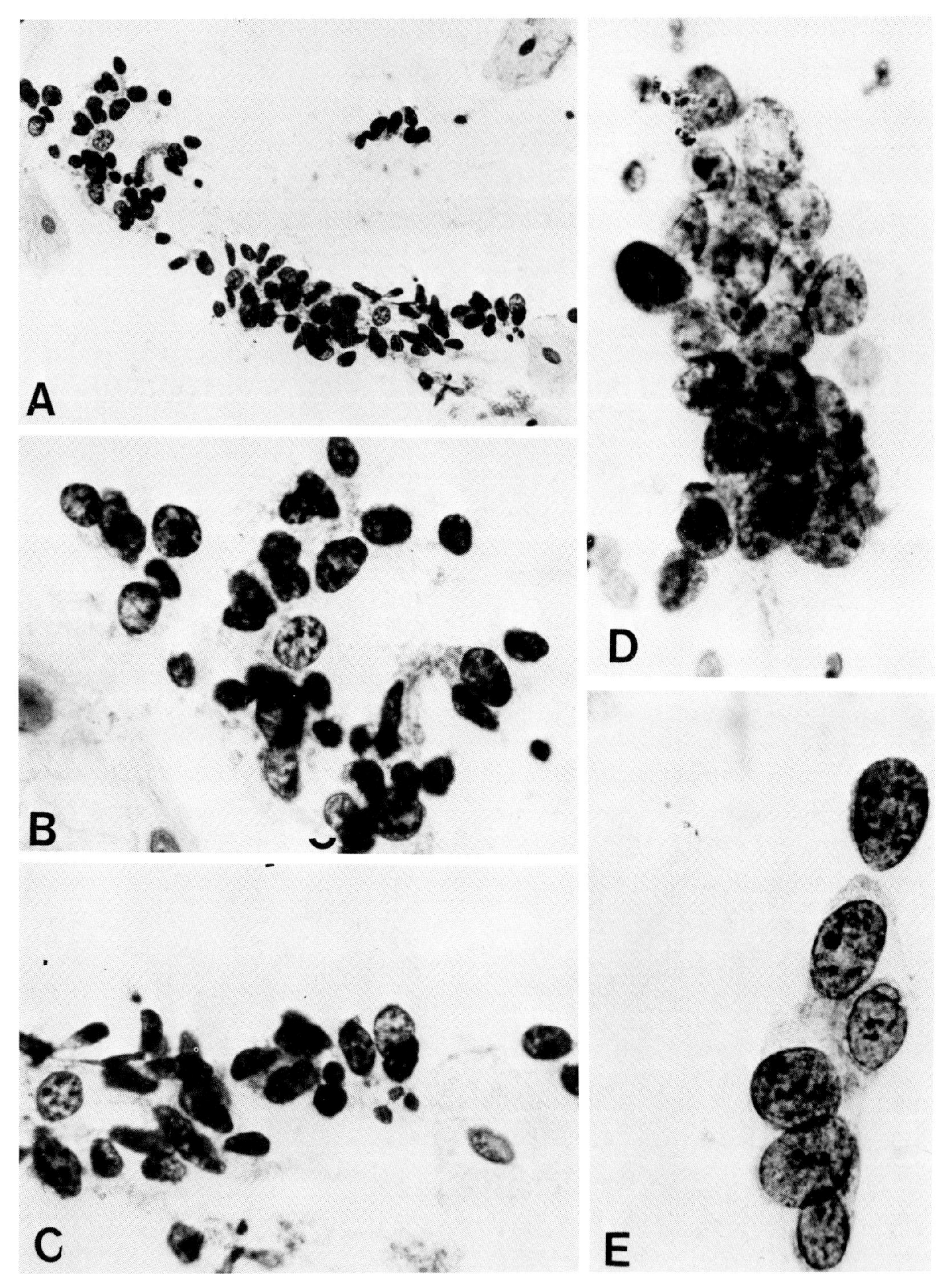

Fig. 3-18. Poorly differentiated cancer cells (*A*) small cell type (×150); (*B,C*) small cell type (×400); (*D,E*) large cell type (×400).

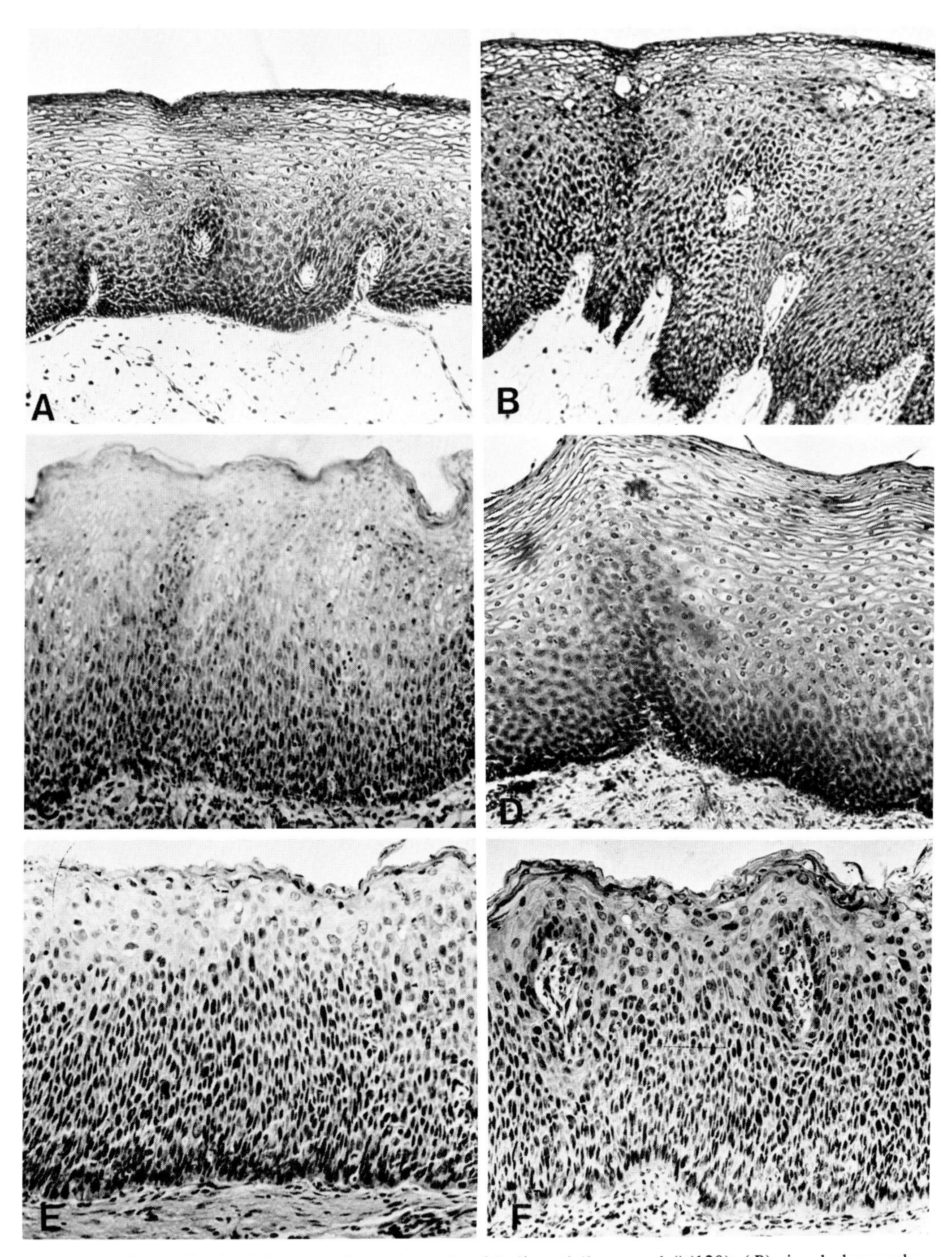

Fig. 3-19. Histologic classification of esophageal epithelium (*A*) normal (×120); (*B*) simple hyperplasia (×120); (*C,D*) mild dysplasia (×120); (*E*) moderate dysplasia (×120); (*F*) severe dysplasia (×120).

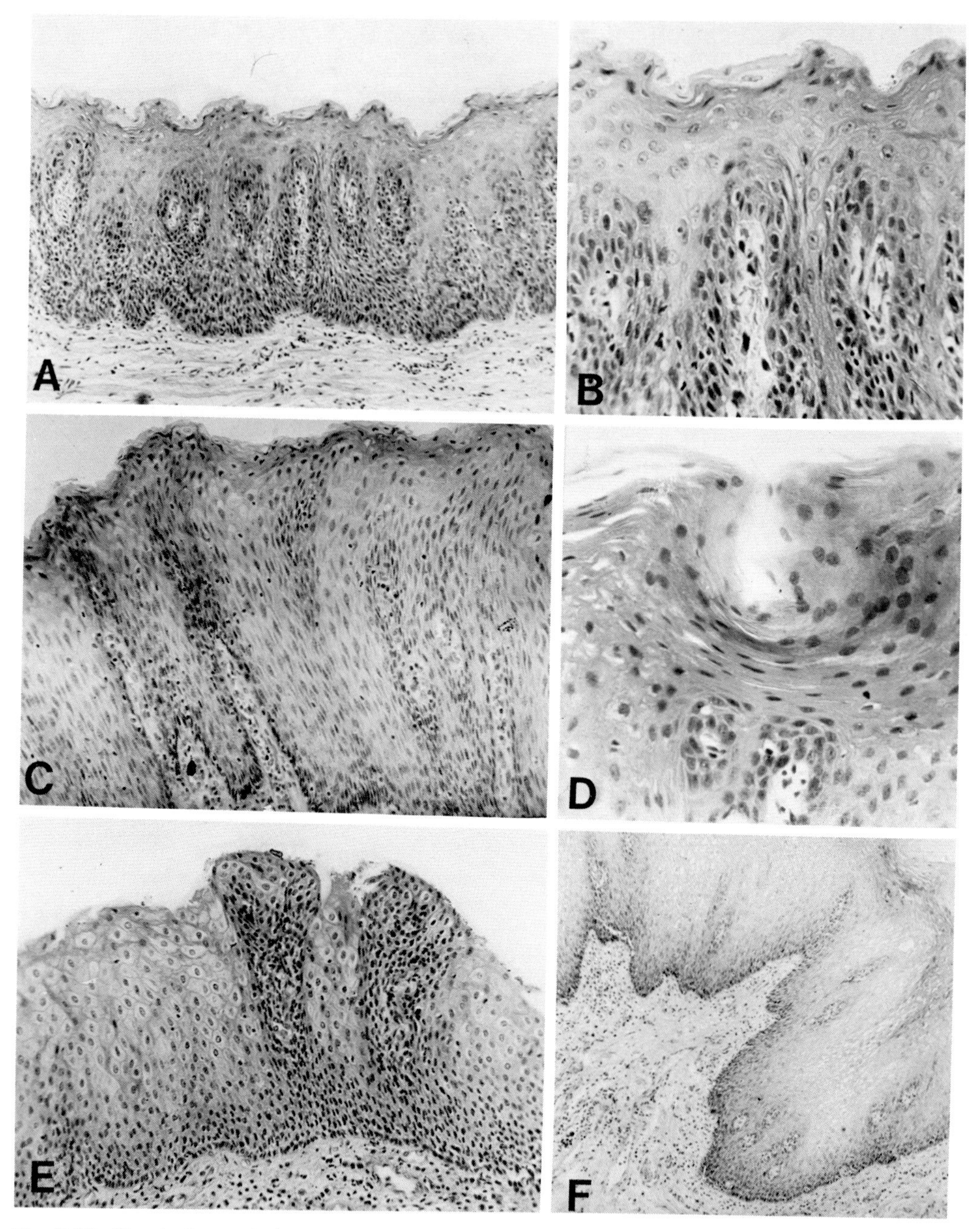

Fig. 3-20. Simple hyperplasia of esophageal epithelium (*A,B*) squamous epithelium is enlarged in thickness. The "papillae" are increased in number and extend into the epithelium (*A:* ×45; *B:* ×180). (*C,D*) Squamous epithelium is enlarged in thickness with parakeratosis (*D:* ×180); the cell appears spindled, but the polarity is normal (*C:* ×90). (*E,F*) Squamous epithelium is thickened (*E:* ×90, *F:* ×45).

dent. Parakeratosis or hyperkeratosis may be observed. The "papillae" are increased in number and reach deeper into the epithelium (Figs. 3-19*B* and 3-20).

Dysplasia: Dysplasia of esophageal epithelium is a defect in normal cell maturation. The morphological features include variation in cytoplasmic maturation, cell configuration, and disturbance of normal nuclear maturation as compared with normal stratified squamous epithelium. There is loss of cell polarity and increased mitotic activity. Nuclear abnormalities include variation in size and shape with hyperchromasia. The thickness of the squamous epithelium may be increased or, less frequently, decreased. Hyperkeratosis or parakeratosis may be observed in dysplasia.

Based on the degree of severity, dysplasia can be divided into three grades:

1. *Mild dysplasia*—Dysplastic changes are confined to the lower third of the epithelium and are similar to basal cell hyperplasia (Figs. 3-19*C* and *D*).
2. *Moderate dysplasia*—The changes are intermediate between mild and severe dysplasia (Fig. 3-19*E*).
3. *Severe dysplasia and carcinoma in situ*—Dysplastic changes involve the entire thickness of the epithelium, which is composed of markedly abnormal cells. However, some maturation may be observed on the epithelial surface (Fig. 3-19*F*).

In prior studies severe dysplasia and carcinoma *in situ* were considered as two separate entities. In view of the follow-up studies on severe dysplasia and its role as a precursor lesion of early invasive cancer, it is now felt that no meaningful and reproducible differentiation can be made between these two entities.

Early Invasive Carcinoma: Malignant cells extend beyond the basement membrane, into the lamina propria or submucosa, but not into the muscularis propria.

CHAPTER

4

The Relationship between Epithelial Dysplasia and Carcinoma of the Esophagus and Its Progression by Cytologic Studies in China

Cytological examination of the esophagus can be used not only in the diagnosis of carcinoma but also in the detection of other pathologic changes, such as inflammation, mucosal epithelial dysplasia, and fungal overgrowth.

During mass surveys for esophageal cancers, many cases of epithelial dysplasia were discovered in addition to early cancers. The rates of both esophageal cancer and esophageal dysplasia were increased in high-incidence areas.[16,29] This observation provides a better understanding of carcinogenesis and also has therapeutic implications. It is hoped that through meticulous management of dysplasia the incidence of esophageal cancer can be reduced.

The development of cancer from epithelial dysplasia and the relationship between the two have been investigated in studies comprised of four steps. First, a comparative study was made of the incidence between dysplasia and carcinoma of the esophagus in various geographic areas; second, a follow-up study was conducted of cases of esophageal dysplasia and cancer (1–12 years); third, severe dysplastic cells in the cancer smear were observed; and fourth, therapeutic studies were made of patients with severe dysplasia of the esophagus.

THE FURTHER RELATIONSHIP BETWEEN EPITHELIAL DYSPLASIA AND CARCINOMA OF THE ESOPHAGUS

Incidence of Dysplasia and Cancer in Different Geographic Areas

The mortality rate of esophageal cancer varies in different areas. It is found that in regions where the rate of cancer is high, the incidence of dysplasia is also high (see Table 4-1).

When Linxian County and Yoatsun Commune were divided into north and south, and Anyang County was divided into east and west along the Wei River, the incidence of esophageal dysplasia coincided with that of esophageal cancer in the different areas (see Table 4-2).

Incidence of Dysplasia and Cancer among Different Age Groups

From 1971 to 1975 a cytologic mass survey was conducted of 54,715 rural inhabitants from high-incidence areas of Henan province. All of those surveyed were over 30 years of age. The survey showed that the incidences of both esophageal dysplasia and cancer rose with increasing age. As displayed in Table 4-3, the incidence of severe

TABLE 4-1
Results of Cytologic Mass Survey in Different Geographic Areas and Nationalities of China

The areas of cytologic mass survey	Mortality of esophageal cancer[a]	Screening survey: Total persons examined	Cancer: No. of cases	Cancer: %	Severe dysplasia: No. of cases	Severe dysplasia: %
(Kazak) Manacy County Sinjiang	240.70	579	11	1.90	33	5.70
Yoatsun Commune Linxian, Henan	169.28	14,002	221	1.58	460	3.25
Shen Guan Commune Linxian, Henan	98.27	19,330	232	1.20	547	2.84
Linxian, Henan	111.00	7,212	78	1.10	135	1.90
Yang Tang Szechuan	79.00	5,697	52	0.91	78	1.39
Zhoung Shang, Hubei (migrants from Henan)	74.00	13,613	60	0.50	136	1.00
Anyang, Henan	71.12	17,471	88	0.50	48	0.27
Min Chong Szechuan	68.34	5,169	27	0.52	16	0.31
Shanghai	10.67	878	0	0	3	0.34
Zhoung Shang, Hubei (non-migrants)	7.68	6,909	3	0.05	16	0.23

[a] Age-adjusted mortality in China per 100,000 people.

dysplasia to that of cancer decreased with age, so that at age 60 it was 1:1. The incidence of cancer increased with age.

Six hundred eighty-one cases from the above study were analyzed. Of the 460 cases of severe dysplasia the average age was 52 years, and of the 221 cases of cancer the average age was 57 years. This suggests a possible relationship between dysplasia and cancer.

Follow-up Studies of Cases with Severe Dysplasia

Five hundred thirty cases with severe esophageal dysplasia have had repeated cy-

TABLE 4-2
Mass Survey of Incidence of Dysplasia and Cancer in Different Geographic Areas

Area	Total persons examined	Cancer: No. of cases	Cancer: %	Severe dysplasia: No. of cases	Severe dysplasia: %
Linxian	7,212				
northern	4,377	57	1.30	98	2.20
southern	2,835	20	0.70	21	0.70
Yoatsun	14,002				
northern	7,769	148	1.92	309	3.80
southern	6,233	73	1.18	151	2.40
Anyang	17,471				
western coast of Wei River	7,081	70	0.98	41	0.57
eastern coast of Wei River	10,390	18	0.17	14	0.13

TABLE 4-3
Comparison of the Incidences of Dysplasia and Carcinoma among Different Age Groups

Age group	No. of persons examined	Cancer		Severe dysplasia		Ratio of severe dysplasia to cancer
		No. of cases	%	No. of cases	%	
30s	17,392	273	1.6	18	0.1	16:1
40s	16,112	268	2.3	116	0.7	3.3:1
50s	11,821	391	3.4	226	1.9	1.8:1
60s	7,080	217	3.1	224	3.2	1:1
70s	1,043	31	2.9	42	4.0	0.1:1
Total	53,448	1,180	2.3	626	1.1	2:1

tological examinations for 1–12 years. As illustrated in Table 4-4, the longer the duration of severe dysplasia, the higher the rate of malignant change.

Results of the Follow-up Studies of Different Grades

Comparisons of different grades of dysplasia and normal cell development showed that the more severe the dysplasia, the greater the malignant potential (see Table 4-5 and Figs. 4-1 to 4-9).

Observation on Severe Dysplastic Cells in the Cancer Smear and the Epithelium Adjacent to Early Esophageal Cancer

An analysis of cyto-smears from 100 cases of early esophageal squamous epithelial cancer showed numerous severe dysplastic cells in 96% of the cases.[30] In 67 cases, the histologic studies of the epithelium adjacent to esophageal cancer showed severe dysplasia in 82% of the cases.[29] The above data suggest that there is an intimate relationship between severe dysplasia and cancer of the esophagus.

TABLE 4-4
Results of Follow-up on 530 Cases with Severe Dysplasia[a]

Time intervals of follow-up (years)	No. of cases of follow-up	No. of cases progressing to carcinoma	Rate (per 100)
>1, < 2	308	26	7.5
2–4	142	29	20.5
5–8	44	15	34.0
9–12	17	9	53.0

[a] Data from Yoatsun Commune of Linxian County.

TABLE 4-5
Comparison of the Incidence of Carcinoma in Different Grades[a]

Cytological diagnosis	Time intervals of follow-up (years)	No. of persons of follow-up	No. of cases with progression to carcinoma (%)
Severe dysplasia	1–12	530	79 (14.8)
Mild dysplasia	1–12	530	5 (1.0)
Normal	1–5	477	0 (0)

[a] $p < 0.001$. Data from Yoatsun Commune of Linxian County.

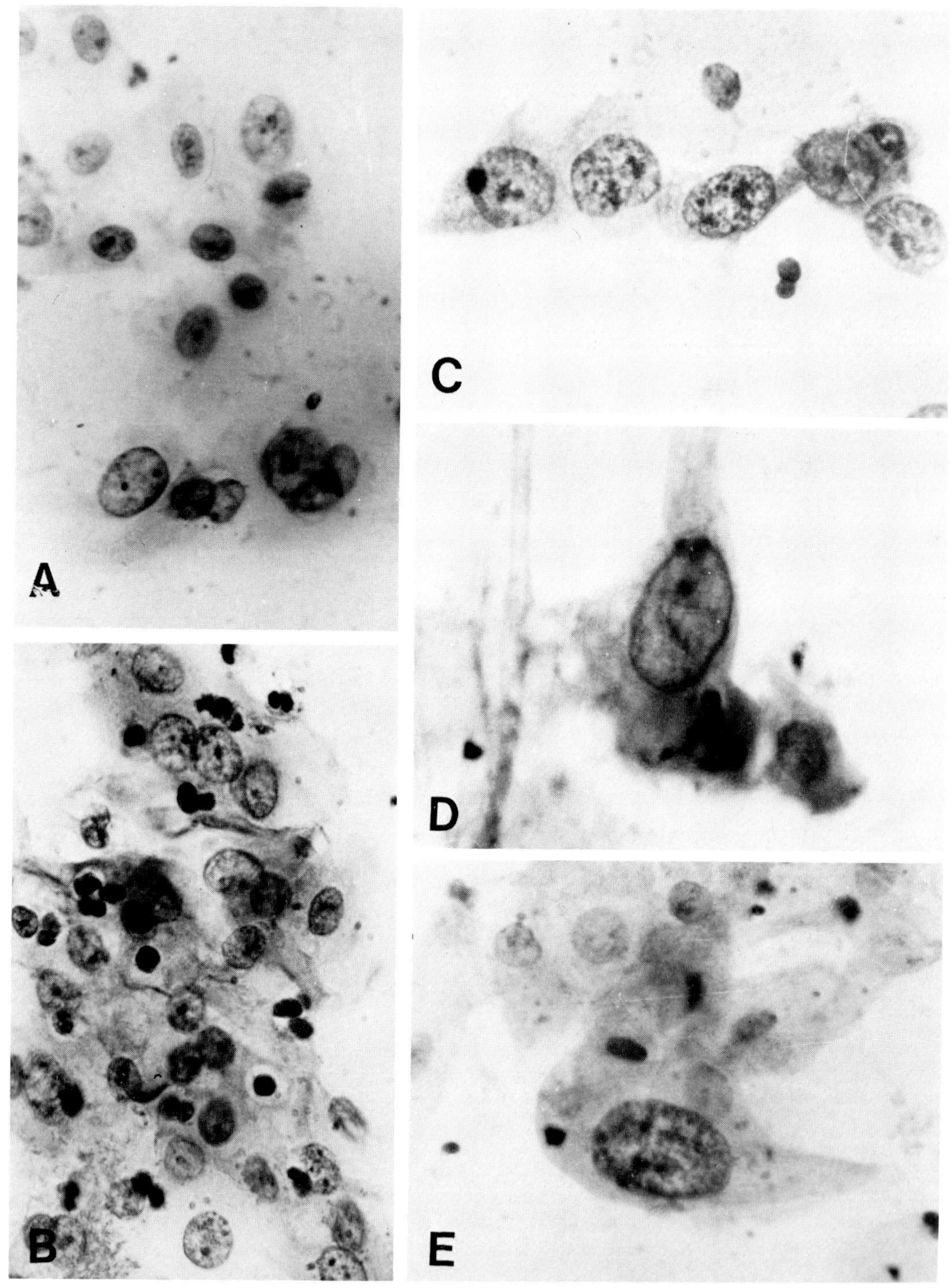

Fig. 4-1. *Case 1.* Severe dysplastic cells (*A,B:* ×200) were found in 1970. Interval cytologic examinations were done four times. The results showed dysplastic cells and cells suspicious for cancer. Cancer cells (*C,D,E:* ×400) were found in 1975.

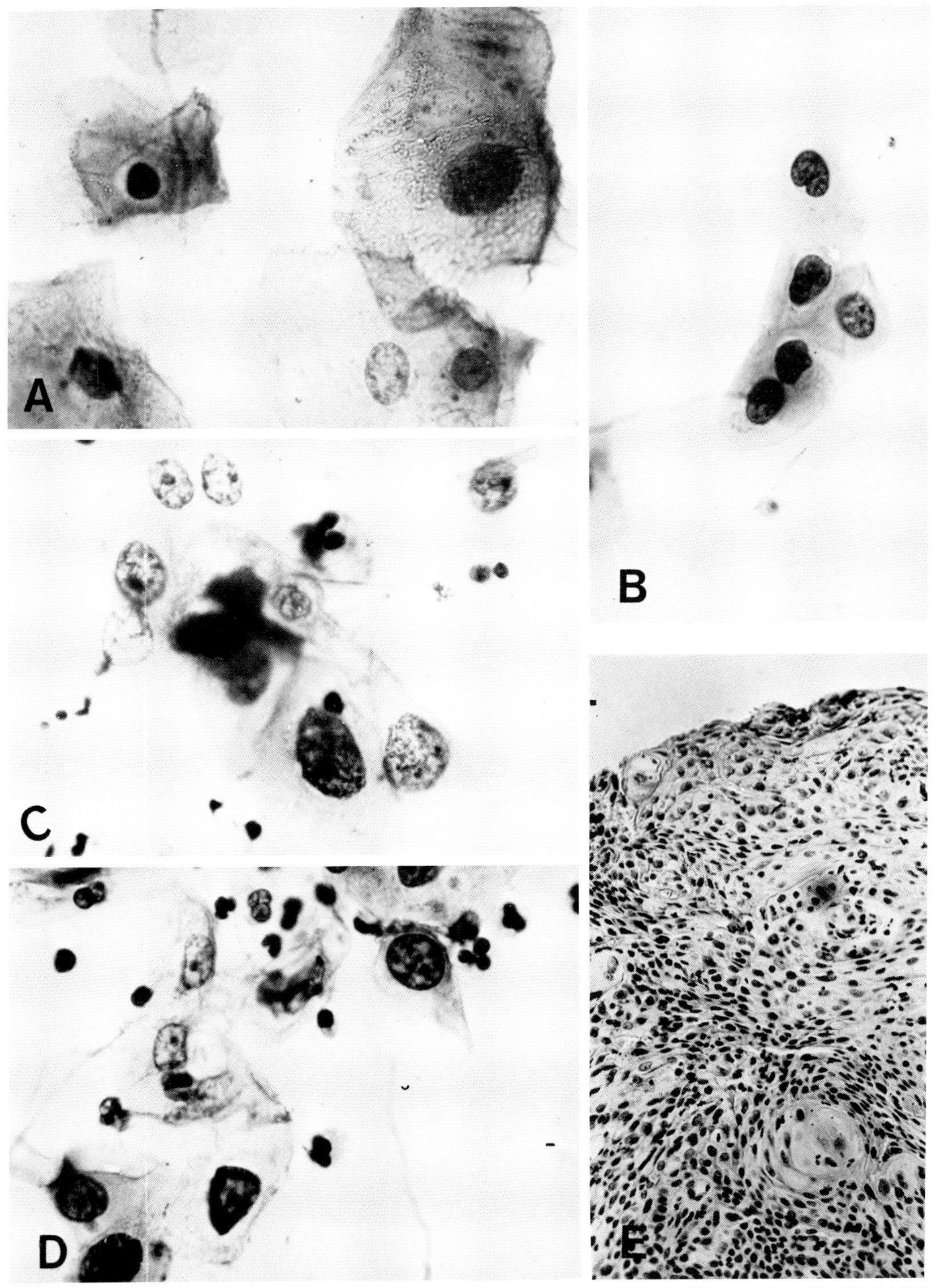

Fig. 4-2. *Case 2.* Dysplastic cells were found in 1974 (*A,B:* ×400). Cancer cells were found in 1977 (*C,D:* ×400). (*E*) Histologic pattern shows early invasive carcinoma (×100).

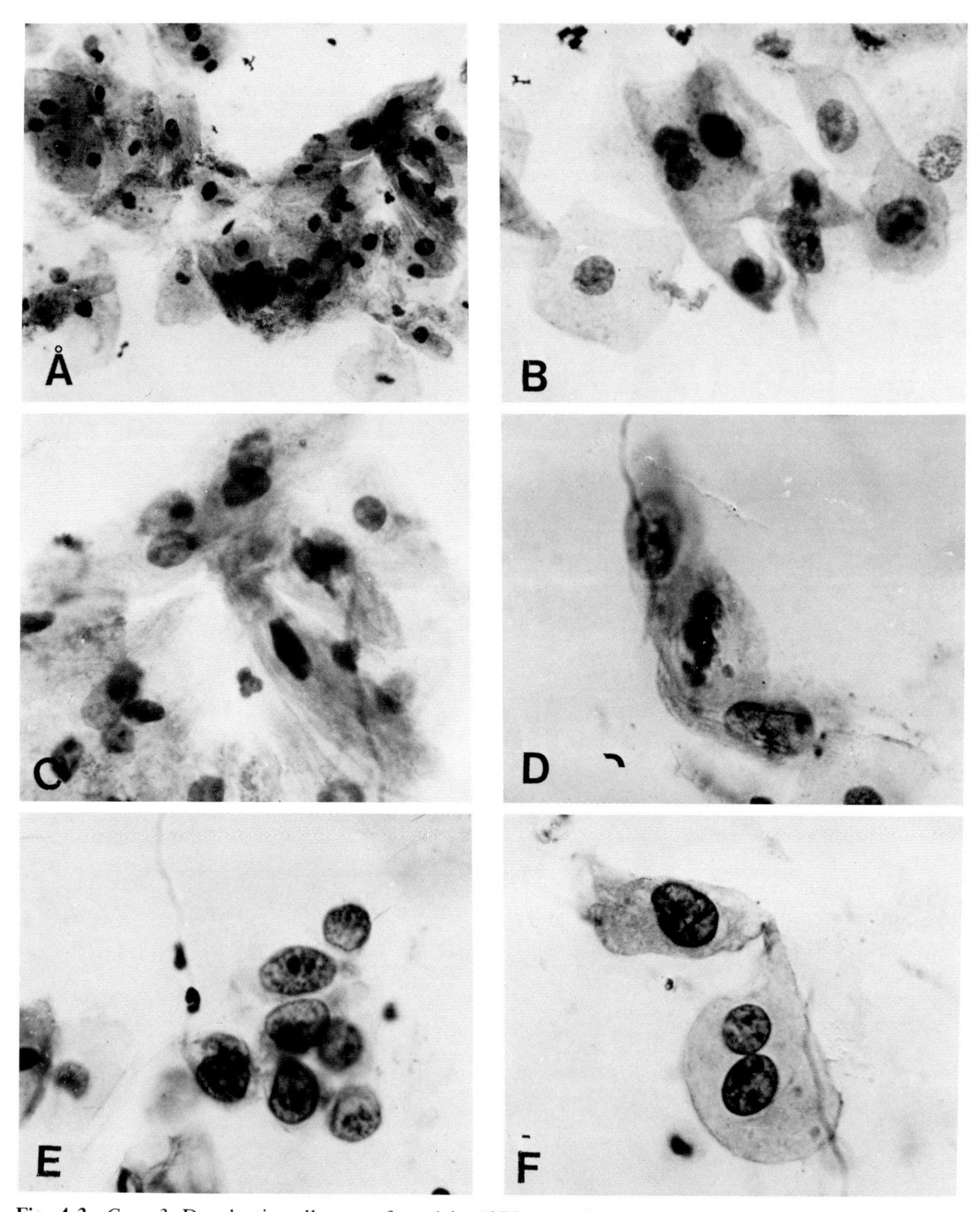

Fig. 4-3. *Case 3.* Dysplastic cells were found in 1975 (*A:* ×90; *B,C:* ×360). Progressed to cancer in 1977 (*D,E,F:* ×360).

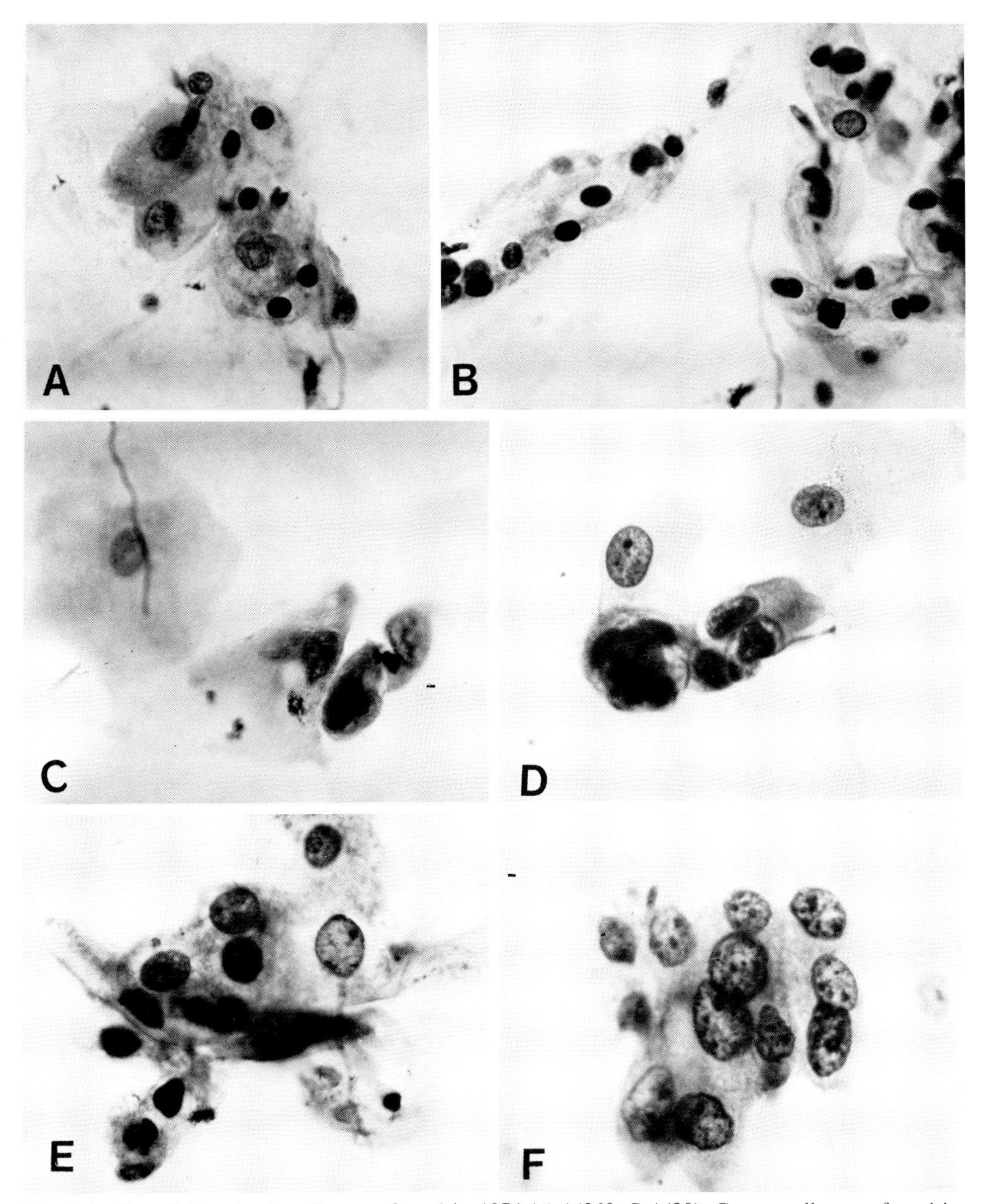

Fig. 4-4. *Case 4.* Dysplastic cells were found in 1974 (*A:* ×360; *B:* ×90). Cancer cells were found in 1977 (*C,D,E,F:* ×360).

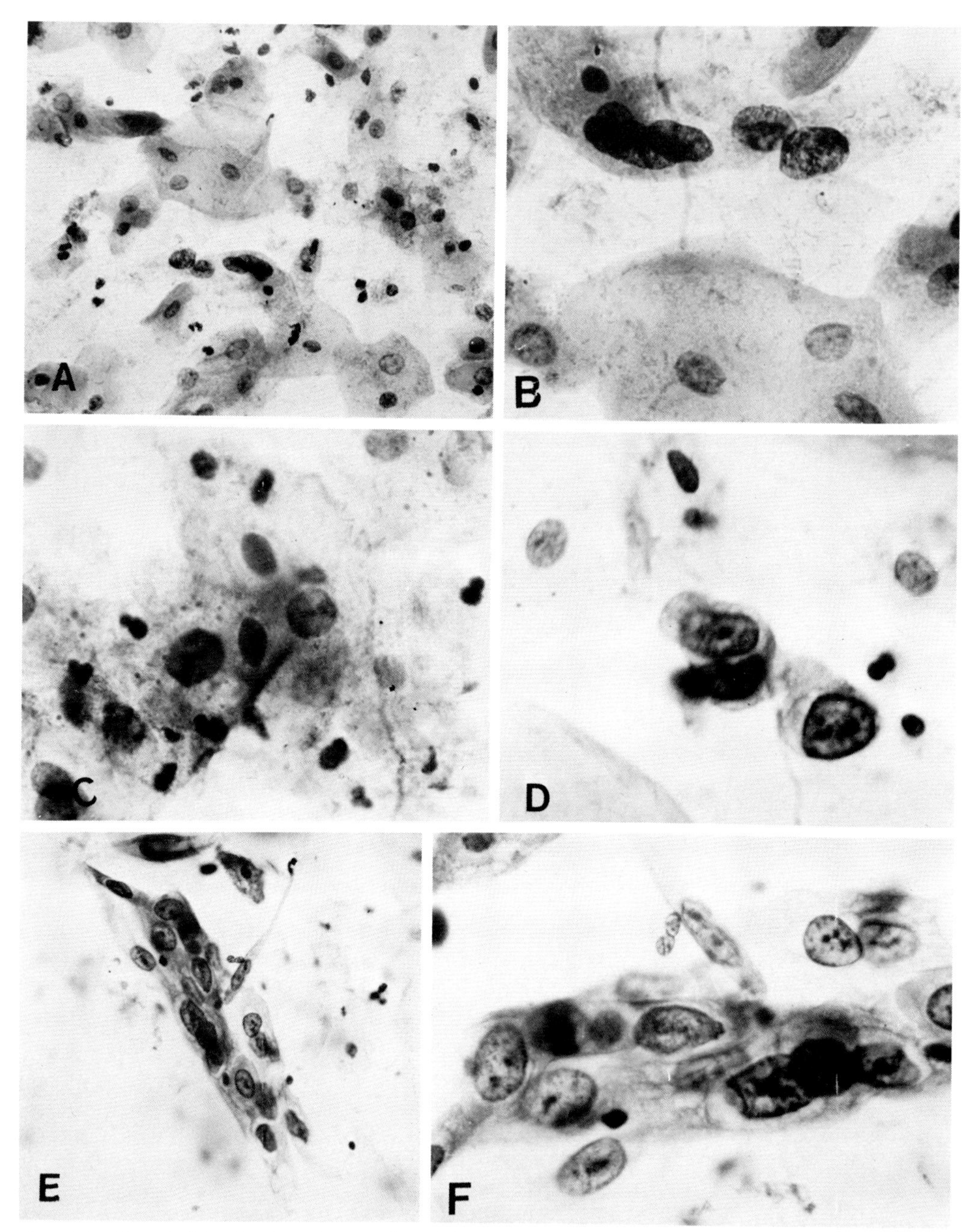

Fig. 4-5. *Case 5.* Dysplastic cells were found in 1973 (*A:* ×90; *B,C:* ×360). Cancer cells were found in 1977 (*D,F:* ×360; *E:* ×135).

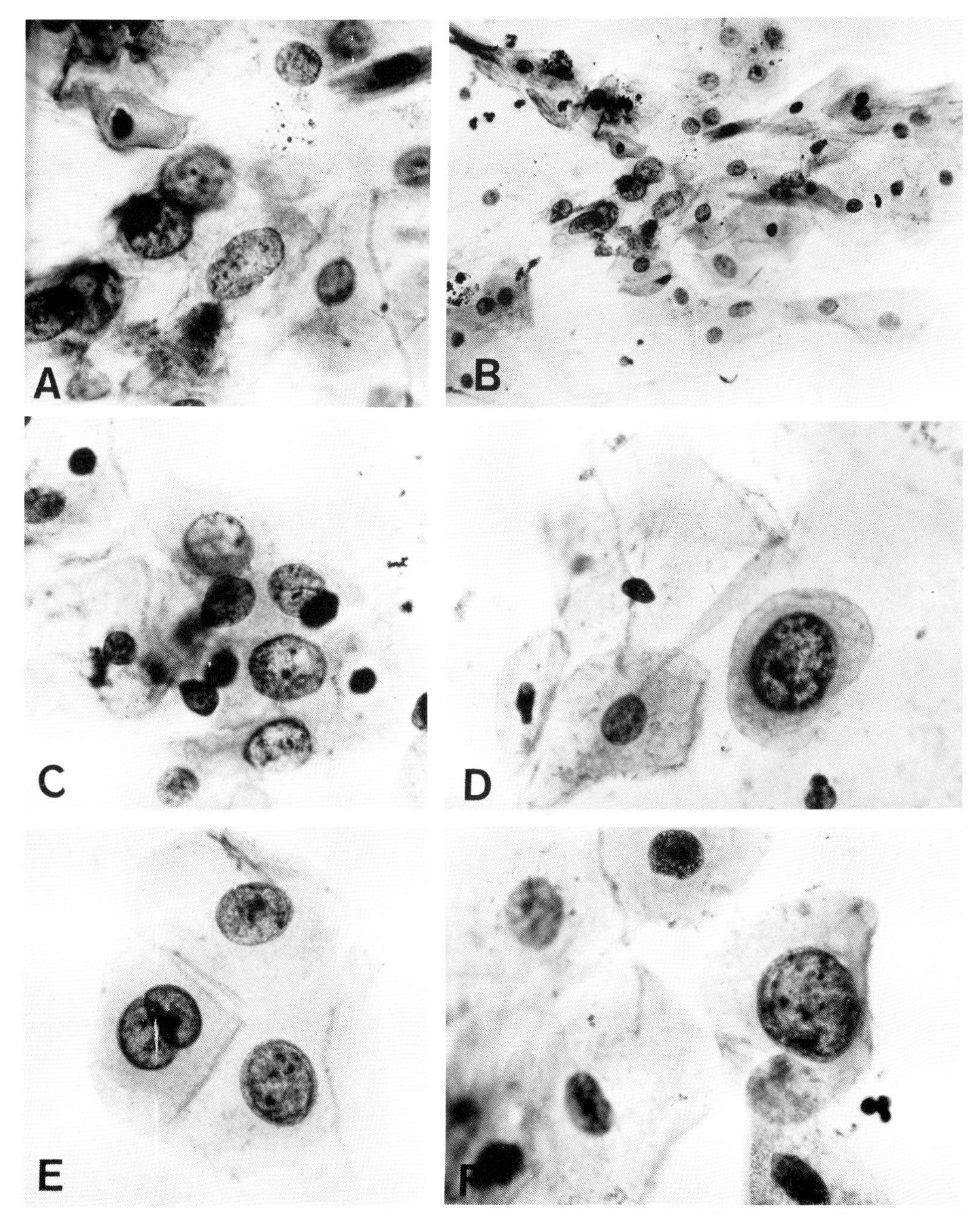

Fig. 4-6. *Case 6.* Dysplastic cells were found in 1973 (*A,C,E:* ×360; *B:* ×90). Cancer cells were found in 1977 (*D,F:* ×360).

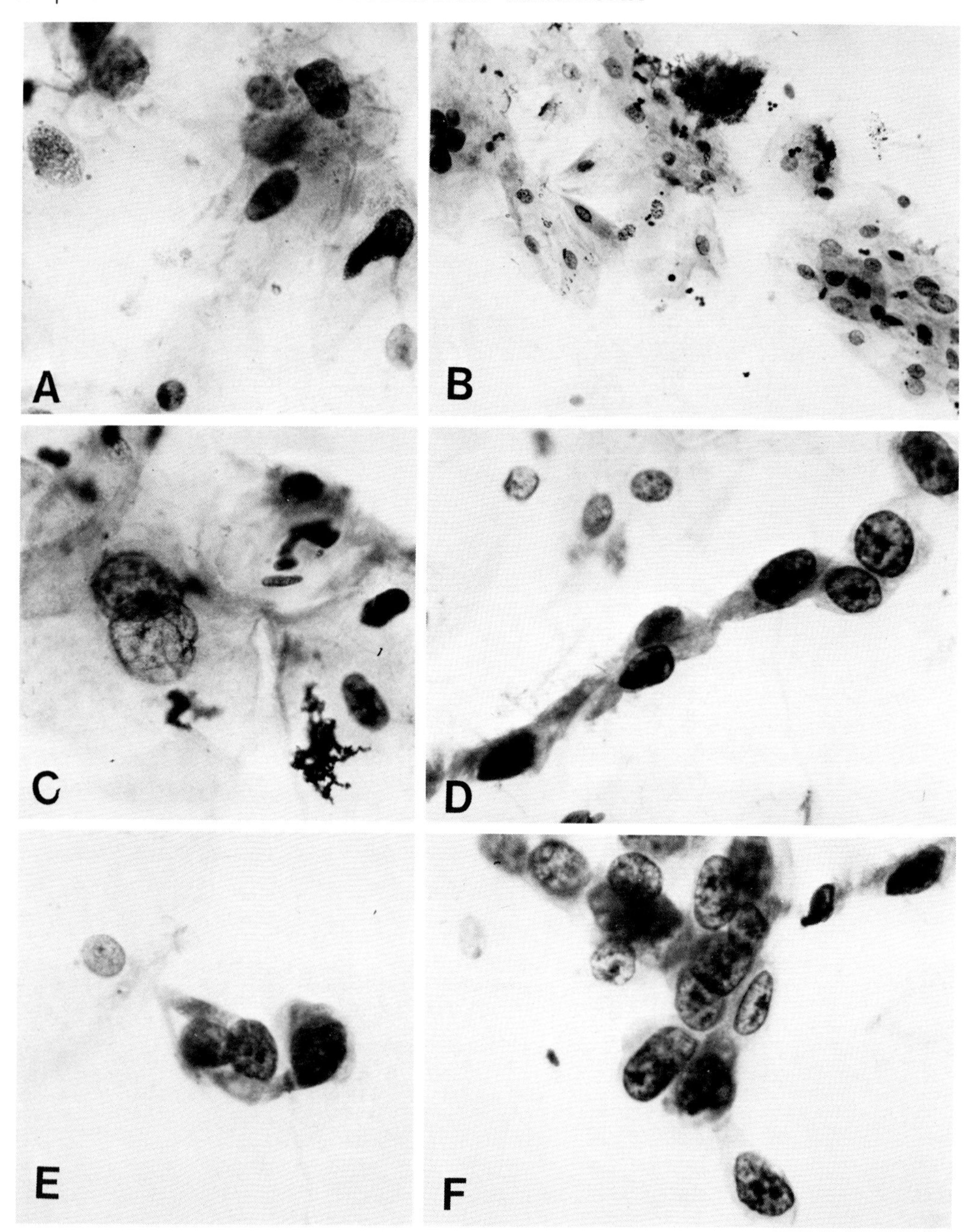

Fig. 4-7. *Case 7.* Dysplastic cells were found in 1974 (*A,C:* ×360; *B:* ×90). Cancer cells were found in 1977 (*D,E,F:* ×360).

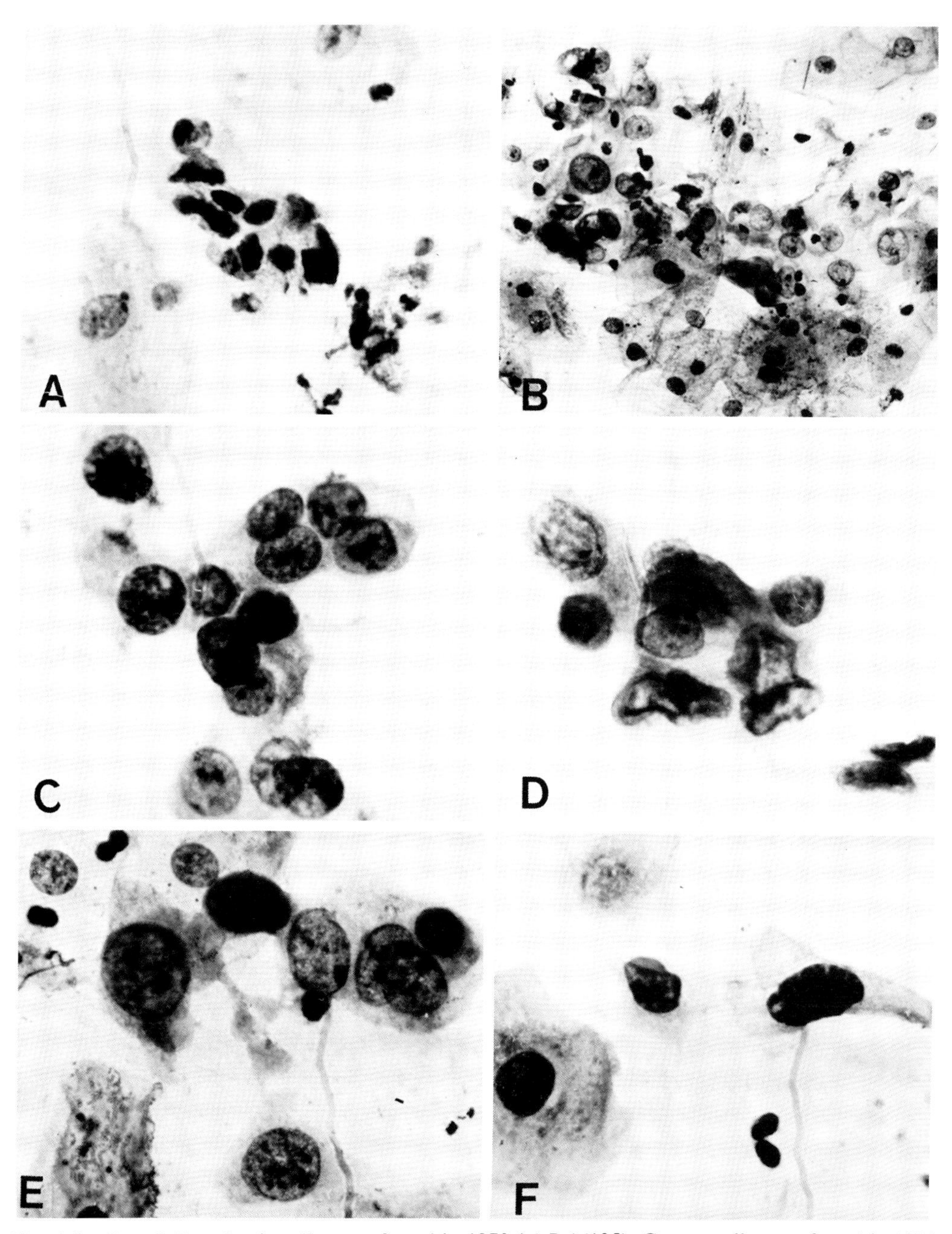

Fig. 4-8. *Case 8.* Dysplastic cells were found in 1970 (*A,B:* ×135). Cancer cells were found in 1974 (*C,D,E,F,* ×360).

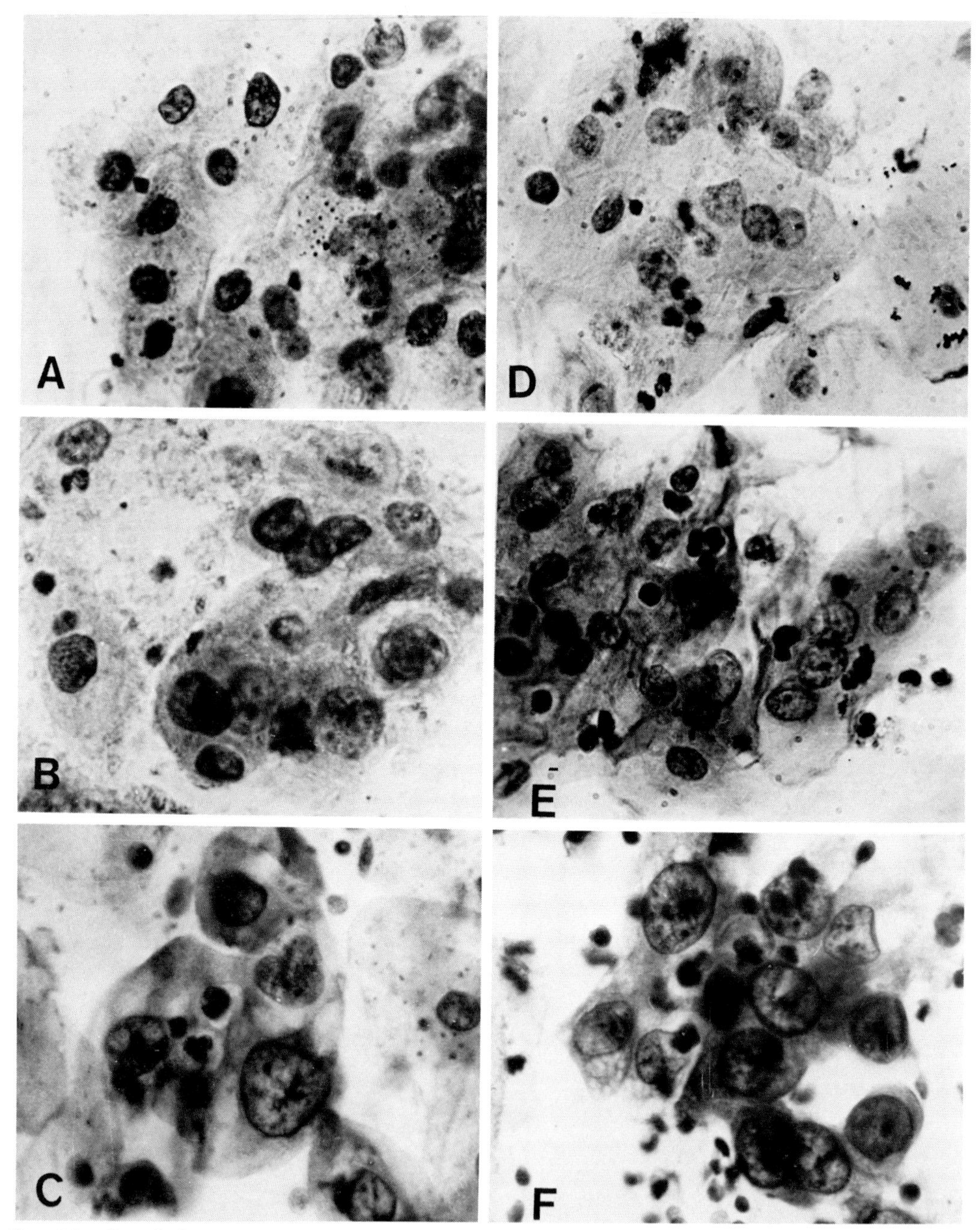

Fig. 4-9. *Case 9.* Severe dysplastic cells were found in 1970 (*A,B:* ×360) and progressed to cancer in 1977 (*C:* ×360).

Case 10. Severe dysplastic cells were found in 1970 (*D,E:* ×360). Cancer cells were found in 1977 (*F:* ×360).

INVESTIGATION OF PROGRESSION OF EPITHELIAL DYSPLASIA AND COMPARISON WITH NORMAL CONTROL[31]

Observations of the behavior of severe dysplasia are recorded in Table 4-6, mild dysplasia in Table 4-7, and normal control in Table 4-8.

LENGTH OF TIME OF PROGRESSION IN SEVERE DYSPLASIA AND CANCER[31]

Progression of Severe Dysplasia to Cancer

Between 1963 and 1977, 112 cases of severe dysplasia that had progressed to cancer were analyzed in Yoatsun and Jinsun Communes of Linxian County. The interval between the diagnosis of severe dysplasia and the first cytologic discovery of cancer cells was determined to be an average of 4 years (Fig. 4-10).

Interval between Early and Advanced Cancer[31]

In China, esophageal cancer is divided into five clinical stages:

1. Stage 0—no radiological evidence of a tumor.
2. Stage 1—evidence of a tumor less than 3 cm in diameter.

TABLE 4-6
Results of Follow-up on 530 Cases with Severe Dysplasia

	Progress to carcinoma		Remaining severe dysplasia		Regressing to mild dysplasia		Regressing to normal	
Follow-up intervals (years)	No. of cases	%	No. of cases	%	No. of cases	%	No. of cases	%
1–12	79	14.91	269	57.00	155	29.00	27	5.20

TABLE 4-7
Results of Follow-up on 530 Cases with Mild Dysplasia[a]

	Progress to carcinoma		Progress to severe dysplasia		Remaining mild dysplasia		Regressing to normal	
Follow-up intervals (years)	No. of cases	%	No. of cases	%	No. of cases	%	No. of cases	%
1–12	5	0.94	62	11.6	283	53.4	180	34.0

[a] Data from Yoatsun Commune of Linxian County.

TABLE 4-8
Results of Follow-up on 477 Normal Cases[a]

	Progress to carcinoma		Progress to severe dysplasia		Progress to mild dysplasia		Remaining normal	
Follow-up intervals (years)	No. of cases	%	No. of cases	%	No. of cases	%	No. of cases	%
1–5	0	0	35	7.3	245	51.4	197	43.0

[a] Data from Yoatsun Commune of Linxian County.

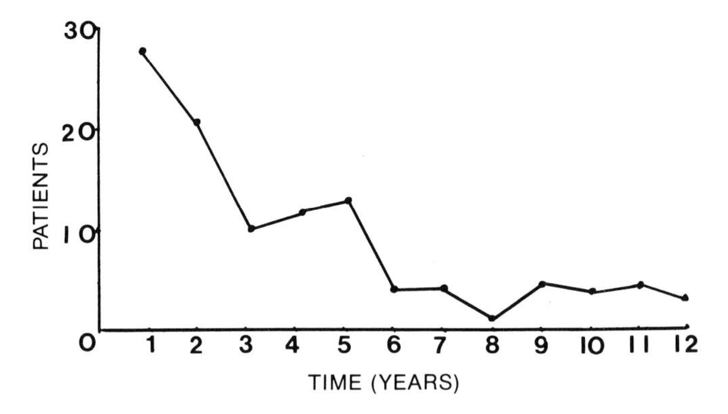

Fig. 4-10. Follow-up study of a group of patients with severe dysplasia that progresses to cancer.

3. Stage 2—evidence of a tumor 3–5 cm in diameter.
4. Stage 3—evidence of a tumor greater than 5 cm in diameter.
5. Stage 4—metastatic disease.

Forty cases of early esophageal cancer were found in a mass survey in Yoatsun Commune in 1974. The aflicted refused treatment and subsequently were observed untreated in 1977. Table 4-9 illustrates the rate of progression for stages 0 and 1.

Stage 0 takes a relatively benign course with 88% remaining at Stage 0, whereas in Stage 1 disease, 67% either progressed to advanced cancer or died.

Progression of Advanced Cancer to Death[31]

Three hundred twenty cases with advanced cancer (Stages 2, 3, and 4) were investigated in Yoatsun Commune from 1974 to 1976. Survival with and without treatment was determined and is shown in Table 4-10.

THERAPEUTIC STUDIES ON PATIENTS WITH DYSPLASIA OF THE ESOPHAGUS

In Linxian County, the incidence of esophageal dysplasia is high, as is the incidence of esophageal carcinoma. Recently we have tried combining Chinese and Western medicine for treatment. The result of therapy was as shown in Table 4–11.[32]

The regression rates subsequent to these treatments were from 48.1 to 68.1%. In the untreated control group the regression rate was 29.8%. Progression to cancer in the control group was 7.4%, and in the treated group from 1.4 to 2.5%. There is a significant difference; however, no therapeutic

TABLE 4-9
The Results of Follow-up on 40 Cases with Early Esophageal Cancer after 3 Years

		Remaining early cancer		Progress to advanced cancer		Died of advanced cancer	
Radiographs	Total no. of cases	No. of cases	%	No. of cases	%	No. of cases	%
Negative (Stage 0)	25	22	88.0	1	4.0	2	8.0
Positive (Stage 1)	15	5	33.3	1	13.3	8	53.4

TABLE 4-10
Average Time of Survival in Advanced Cancer

	No. of cases	Average time of survival in months
Treated	199	15.9
Untreated	121	8.5
Total	320	12.2

TABLE 4-11
Cytologic Examinations of Patients with Severe Dysplasia of Esophageal Epithelium after Drug Therapy

		Regressing to mild dysplasia or to normal		Remaining mild dysplasia		Progressing to cancer	
Group	Total No. of cases	No. of cases	%	No. of cases	%	No. of cases	%
Untreated control	215	64	29.8	135	62.8	16	7.4
Antitumor B III (1976–1977)	162	78	48.1[a]	80	49.4	4	2.5[b]
Antitumor B (1977–1978)	72	54	75.0[a]	17	23.6	1	1.4[b]
Tilorone	43	21	48.8[a]	21	48.8	1	2.3[b]

[a] $p < 0.01$.
[b] $p > 0.05$.

effect was noted in certain patients with second-degree severe dysplasia or suspected cancer.

CONCLUSION

This chapter has delineated the relationship between esophageal dysplasia and carcinoma of the esophagus with the use of material gathered from cytologic mass surveys. It was found that where there were high incidences of and mortality rates from esophageal carcinoma, the same was true for dysplasia. A close relationship between dysplasia and carcinoma was assumed and then confirmed by analysis of the data. Age studies found that severe dysplasia preceded carcinoma by 5 years and severe dysplasia was thought to be the precursor lesion. Follow-up studies showed that with increased duration of severe dysplasia, the rate of malignant conversion also increased; and that with more severely dysplastic lesions, the malignant potential is greater (severe dysplasia to carcinoma at a rate of about 15 times greater than mild dysplasia). These results supported the belief that severe dysplasia of the esophagus is the precancerous lesion.

Cancerous changes of esophageal epithelium run the spectrum from mild dysplasia to severe dysplasia to cancer. Although lesions generally develop into cancer, regressions do occur from severe to mild dysplasia. The average time for severe dysplasia to progress to early cancer is 4 years; from early cancer to advanced cancer 1 year for stage 1 and more than 3 years for Stage 0; and advanced cancer to death 1 year (see scheme).

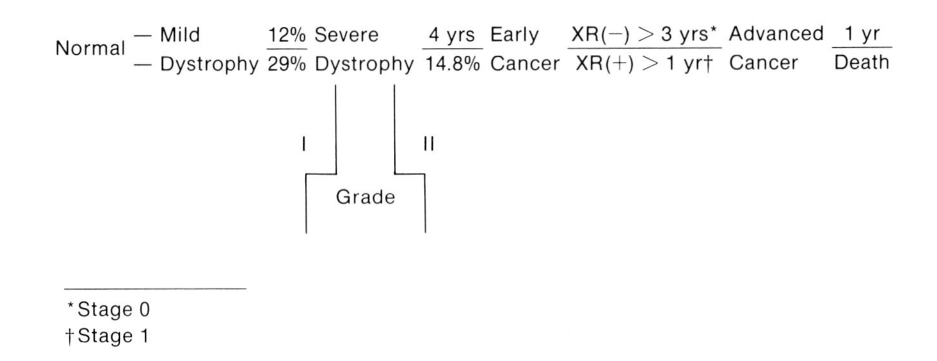

Since 1974, patients with severe dysplasia have been treated and, although we have some good results, further work is needed before any conclusions can be drawn.

CHAPTER 5 Morphologic Characteristics of Dysplasia

Dysplasia of the esophagus is reversible, nonmalignant, and therefore, presumably, a controlled cellular abnormality. When the inciting stimulus is removed, dysplasia returns to normal. Dysplasia is a defect in the normal cell maturation pattern which is manifested morphologically by variations in cytoplasmic and normal nuclear maturation as compared with normal stratified squamous epithelia of the esophagus. There is a loss in the uniformity of shape and arrangement of the individual cells, as well as a loss in their architectural orientation. Frequently, severe dysplasia is a forerunner of cancer.

GROSS MORPHOLOGY OF ESOPHAGEAL DYSPLASIA

In the fresh specimen, the dysplastic epithelium is grayish-white and is thicker than normal. After fixation, the mucous membrane is wrinkled and appears coarse and thick.

HISTOLOGY OF ESOPHAGEAL DYSPLASIA

When studied microscopically, the chief findings are:

1. The squamous epithelium may be increased in thickness from as few as five layers to as many as 30 layers or more.
2. The epithelial cells are increased in number. The morphologic features include variation in cytoplasmic maturation. Dysplastic cells exhibit considerable pleomorphism (variation in size and shape) and frequently have scanty cytoplasm. Cellular arrangement is disorderly (loss of polarity). The nucleus is oval, spindle or pleomorphic in shape, and often is hyperchromatic and abnormally large. Mitotic figures are more abundant than usual and are frequently found in abnormal locations within the epithelium.
3. A normal maturation pattern, hyperkeratosis, parakeratosis, and dyskeratosis in the superficial epithelium may be observed.

CYTOLOGIC OBSERVATIONS ON SEVERE ESOPHAGEAL DYSPLASIA

Follow-up studies of esophageal cytology showed a continuous transition from normal esophageal epithelium to atypia, mild dysplasia, severe dysplasia, and carcinoma. As mentioned earlier, for patients with severe dysplasia there is a high risk of progression to esophageal carcinoma. Mild dysplasia, however, has not been shown to undergo obvious malignant transformation; therefore, the following discussion will be confined to the cytologic characteristics of severe dysplasia.

Cellular characteristics of severe dysplasia were established by analyzing the cell populations in samples from 430 patients with severe dysplasia (one-third of carcinomas *in situ*) and 100 patients with early cancer (two-thirds of early invasive carcinomas).

Cellular Types

According to morphologic characteristics, epithelial dysplastic cells may be divided into three types: superficial, intermediate, and parabasal.

The frequency of the distribution of predominant cell types was determined for 430 cases of severe dysplasia and 100 cases of early carcinoma, as shown in Figure 5-4. It may be noted that there was an increase in the proportion of small, abnormal cells in cancer as compared with dysplasia. Thus, the smear patterns reflect the level of histologic abnormality.

As a general rule, if the dysplastic cells are predominantly of the superficial type, there is less likelihood of a tumor. Conversely, if the dysplastic cells are parabasal and numerous, the probability of carcinoma is high.

Cellular Shape

The shape of an abnormal cell may also reflect the maturity of the parent tissue. Normal mature squamous cells are polygonal in form and vary in size. In severe dysplasia, the majority of the cells are of normal size and shape, but occasionally large and small types of superficial cells, with deeply eosinophilic cytoplasm, are seen. "Fiber" cells were noted in 3.9% of the 430 cases of severe dysplasia and in 10% of the 100 cases of early cancer. These cells were usually associated with dyskeratosis or parakeratosis of the esophageal epithelium. Other irregularly shaped cells of squamous type are commonly observed in dysplastic and early cancer samples. The tadpole shape has never been observed in the study of dysplasia, but it has been found in advanced cancer. Round or oval cells with deeply eosinophilic, abundant cytoplasm were also seen in dysplasia and early cancer. The nuclei of these cells were pyknotic and either uniform or convoluted in shape.

Nuclear Morphology

Nuclear morphology was investigated in histologic sections of 18 cases of severe dysplasia by use of the oil immersion lens. A total of 3,550 abnormal cells were recorded. The nuclei were oval or round in 87.7% and irregular in 12.3%; the chromatin pattern was finely granular in 56.7%; chromatin distribution was found in 1.2% of the cells; and only 0.1% had an uneven distribution with an irregular nucleus. There was a thick nuclear rim in 82.5% of the cells.[33]

Nuclear morphology was also studied in the cytosmears. Dysplastic cells from the 430 cases with severe dysplasia and 100 cases of early cancer were studied. The morphologic changes in the nuclei were as follows:

1. *Nuclear size.* In dysplastic superficial cells, the majority of the nuclei were greater than 15 μm in diameter (normal 6–8 μm). However, the second most common sighting was cells with small nuclei and a coarse pattern of chromatin in most of the dysplastic smears.
2. *Nuclear shape.* The majority of the nuclei were round or oval; few were irregular. The population of dysplastic cells with irregular nuclei differed from severe dysplasia to cancer. In severe dysplasia, 3% of the cases showed superficial dysplastic cells with irregular nuclei; 2.3% had intermediate cells; and 1.2% showed parabasal dysplastic cells. In cancer, 31.2% of the cases had superficial dysplastic cells; 37.5% had intermediate cells; and 41.2% had parabasal dysplastic cells. Usually the dysplastic parabasal cells associated with

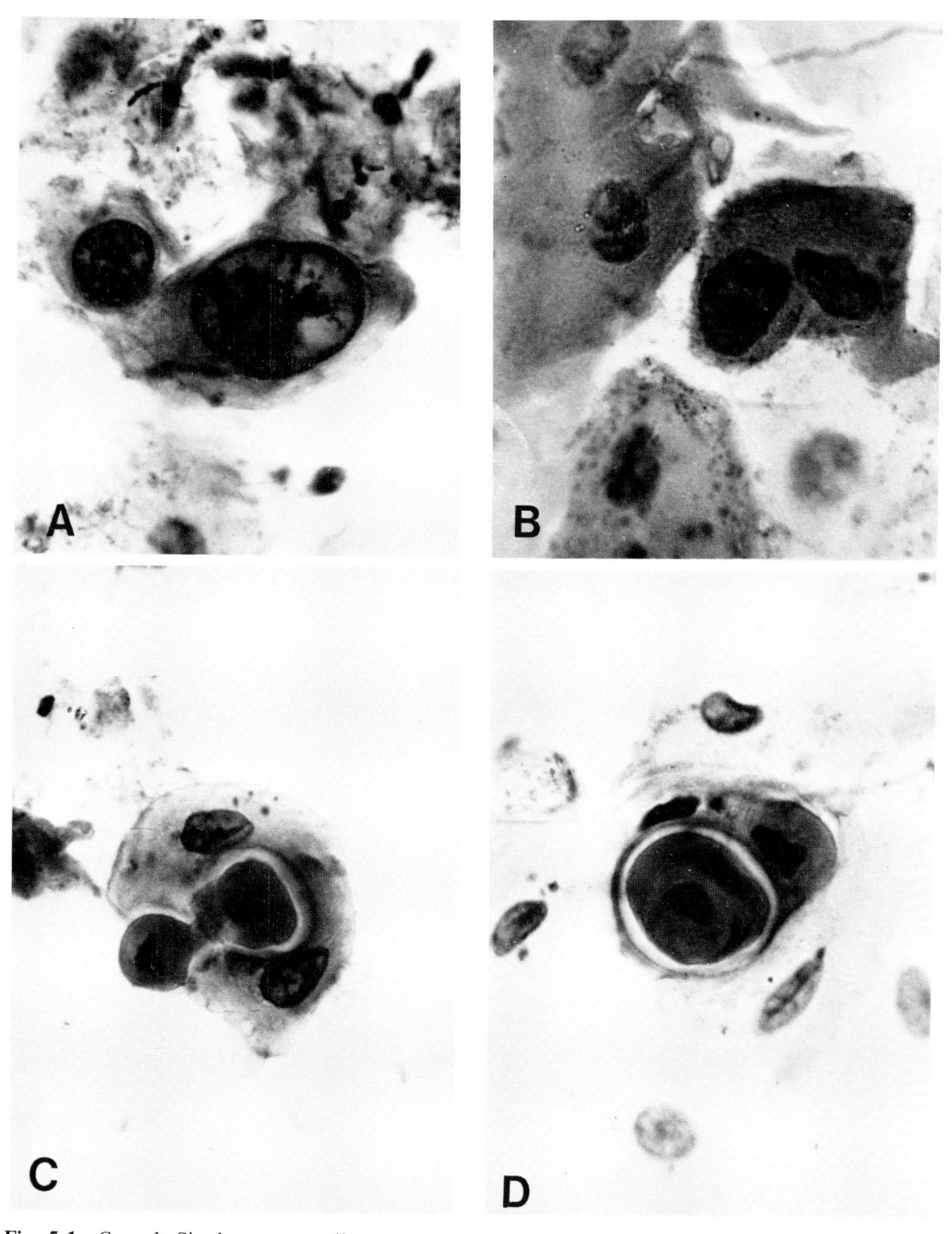

Fig. 5-1. *Case 1.* Single cancer cells were detected in mass screening (*A:* ×540), squamous cell pearls (*C,D:* ×540), and severe dysplastic cells (*B:* ×540). Two separate endoscopic examinations were negative. No cancer could be found after 5 years.

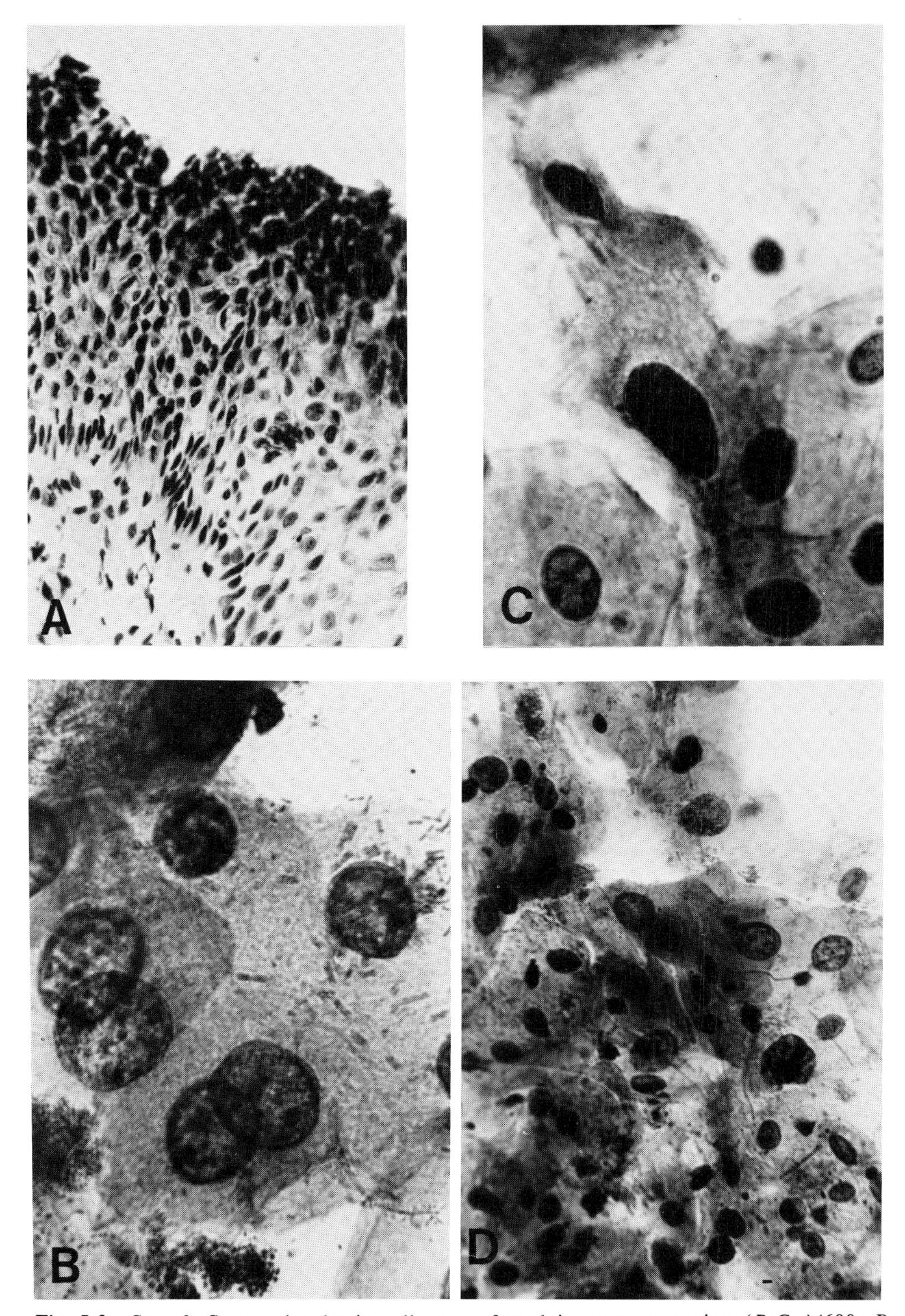

Fig. 5-2. *Case 2:* Severe dysplastic cells were found in mass screening (*B,C:* ×600; *D:* ×200). Findings were confirmed by histologic section (*A:* ×200).

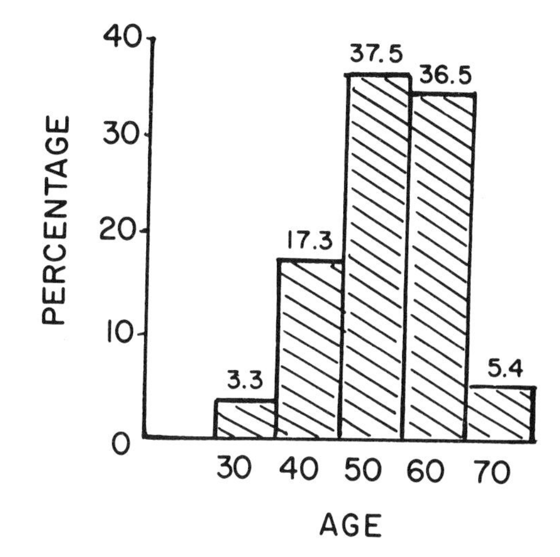

Fig. 5-3. Distribution of various age groups in severe dysplasia of the esophagus.

cancer may be derived from clumps of cancer cells or from dysplastic epithelium. It is evident that in the presence of dysplastic parabasal cells and dysplastic cells with irregular nuclei, the likelihood of cancer is high.

3. *Nuclear structures.* In most cases, the chromatin was granular and uniformly dispersed, but occasionally it was coarsely granular. The nuclear membrane was uniform in thickness. Nucleoli were rarely seen; however, occasionally one might be encountered in a basal cell.
4. Double or multiple nuclei were infrequently seen in dysplastic cells.

Nucleo–Cytoplasmic (N:C) Ratio

An increase in the ratio of the diameter of the nucleus to the diameter of the cytoplasm (N:C) is a valuable criterion of dysplastic or

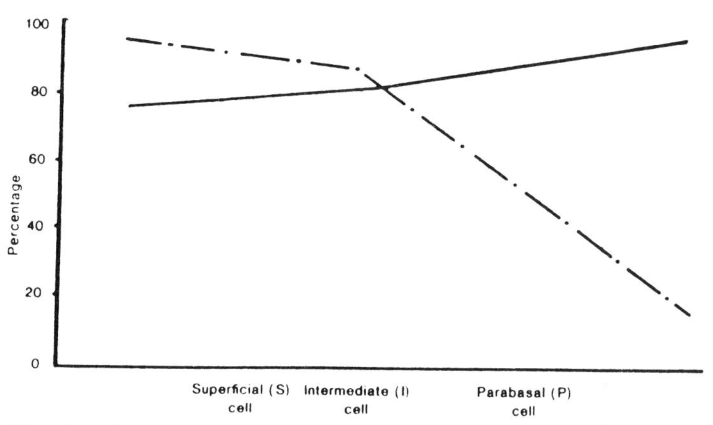

Fig. 5-4. The distribution of various types of abnormal cells in severe dysplasia and early cancer. The increase in small abnormal cells is characteristic of carcinoma. Dysplasia group: S > I > P (broken line); cancer group: P > I > S (solid line). S = superficial; I = intermediate; P = parabasal. (From Koss and Coleman: *Advances in Clinical Cytology, Vol. 2,* Masson Publishing USA, Inc., New York, 1984; reproduced with permission.)

malignant cells. The N:C ratio increases with the diminishing size of the abnormal cell. An extremely high N:C ratio is considered characteristic of cell samples from *in situ* squamous cell carcinoma. The N:C ratio in parabasal cells was investigated in smears from 100 patients with early cancer. A marked increase in the N:C ratio was observed in 28% of the cases; a moderate increase in 39%; and a slight increase in all other cases.

The Smear Background

In abnormal cells derived from a dysplastic reaction alone, most of the cellular samples presented with various degrees of infection. Of the 440 cases with severe dysplasia investigated, 47.5% showed mild infection, 27.6% showed moderate infection, and 11.5% showed severe infection. Seventy-two percent of the samples revealed fungus. The amount of fungus increased with the degree of infection. There is generally no evidence of tissue destruction or red blood cell breakdown. It may be of interest that in 62 esophageal smears obtained from patients at Montefiore Hospital in New York, there was no evidence of fungal infection.

SUMMARY OF CYTOLOGIC ABNORMALITIES THAT REQUIRE IMMEDIATE FOLLOW-UP OF PATIENTS (Figs. 5-1 and 5-2)

1. Presence of dysplastic cells of parabasal type.
2. Presence of dysplastic cells with irregular nuclei.
3. Presence of pleomorphic dysplastic cells, i.e., fiber or spindle forms or other irregular forms.
4. Anisocytosis and anasokaryosis with deep eosinophilic staining reaction in the cytoplasm.
5. Karyomegaly or irregular micronuclei.

MICROSPECTROPHOTOMETRY (MSP)

In 1978, Lin and associates reported the microspectrophotometry (MSP) measurement of DNA content in the nuclei of esophageal epithelial cells.[34] The DNA content of the nuclei of severe dysplastic cells was more than that of normal and mild dysplastic cells, but was less than that in the carcinoma cells. Thus, our cytologic diagnosis correlated with the MSP measurement data.

AGE AT DETECTION

In cases based on cellular evidence of severe dysplasia, the mean age of detection of 460 consecutive cases with dysplasia was 51.7 years (51.00 in male and 52.51 in female). The distribution of various age groups is shown in Figure 5-3.[38]

The mean age of detection of 221 consecutive cases with cancer was 57.3 years. The cancer group is 5.6 years older than the severe dysplasia group. In the 30-year age group, 20.4% showed severe dysplasia and 3.2% showed cancer. In the 60-year age group, 17.2% showed severe dysplasia, 39.1% showed cancer. As illustrated in the figure,

TABLE 5-1
The Relationship between Cytodiagnosis and Microspectrophotometry

Group	No. of cases	DNA content mean	Hexaploid		Octaploid		Aneuploid	
			No. of cases	% of cases	No. of cases	% of cases	No. of cases	% of cases
Normal	25	26.08	28.0	1.4	0	0	0	0
Mild dysplasia	25	17.14	36.0	2.0	8.0	0.16	8	0.16
Severe dysplasia	100	39.02	97.0	17.6	39.0	1.5	12	0.4
Cancer	50	67.79	98.0	45.2	98.0	39.1	96	19.6

the age of detection of cases with severe dysplasia is younger than those with cancer.

SYMPTOMS OF SEVERE DYSPLASIA

Seventy-five percent of the cases of severe dysplasia were without symptoms. Three and three-tenths percent had difficulty swallowing, 7.4% had a feeling of a foreign body when swallowing, 16.5% felt discomfort or a burning sensation when swallowing, 2.8% had pain behind the sternal bone or soreness in the back, 7.8% belched when eating, 1.8% vomitted mucus.[38]

CHAPTER 6

Early Esophageal Squamous Cell Carcinoma

The extensive progress in esophageal cancer detection by cytologic techniques described in this chapter began in areas with high incidence from 1970 to the present. Individuals over 30 years of age were invited to participate in mass screening surveys conducted at 4-year intervals. If dysplasia was discovered, the patient was examined yearly. During these surveys, over 80% of the eligible population in the area were examined.

From 1970 to 1975 the author participated in several detection programs encompassing 81,187 asymptomatic people age 30 and over. Eight hundred eighty carcinomas of the esophagus were detected, of which 649 (73.75%) were classified as early cancer (carcinoma *in situ* and superficially invasive carcinoma). The rate of early cancers varied from a low of 56% in Anyang Region to a high of 85% in Shen Guan Commune.

The remarkable figures for early carcinoma detected by cytology must be compared with figures for symptomatic patients seen in hospitals. Among 35,440 patients, 11,108 cases of esophageal cancer were diagnosed. Only 978 (8.8%) could be considered early cancer. It should be pointed out that in the Cancer Institute (C.A.M.S.) in Peking the rate of early esophageal cancer was only 0.23%.

The comparison of survival rates for patients with early carcinoma treated by surgery to patients with advanced cancer treated by surgery is striking. Of 170 patients with early carcinoma, 90.3% survived for 5 years. The survival rate for advanced carcinoma was only 29%.

CLINICOPATHOLOGIC APPEARANCE OF EARLY ESOPHAGEAL CANCER

Early esophageal cancer may be defined as follows:

1. The lesion is confined to the mucosa and the submucosa without infiltration into the muscular layer; no metastases are present.
2. The length of the lesion is less than 3 cm by radiographic studies.
3. On endoscopy, the lesion is limited in the mucosa. Erosions, plaques, coarse thickening of the epithelium, vasodilatation, or mucosal folding can be seen. Masses or ulcerations are not present.

CLASSIFICATION OF THE GROSS SPECIMEN IN EARLY ESOPHAGEAL CANCER

Most of the esophageal cancers reported in China and elsewhere are already moderately advanced or late. In recent years, many early esophageal cancers have been discovered by using the cytologic mass survey in high-incidence areas. The gross features of early cancer are quite different from those of advanced carcinoma.

The Cancer Prevention and Treatment Team in Henan classified early esophageal cancer as three types: 1) flattened type, 2) mild depression type, and 3) mild elevation type. Microscopically, early esophageal cancer is classified as carcinoma *in situ* and early infiltrating cancer. In some cases, invasive carcinoma may be derived from the basal layers of mildly dysplastic epithelium. From the 100 early cases discovered in Linxian County, the Coordinating Group on Research of Esophageal Cancer, Chinese Academy of Medical Sciences suggested the following classification[35]:

Occult Type

The lesion, being thin and flat, can barely be detected by the naked eye. In fresh specimens, the diseased area shows only a slight deepening in coloration on the mucosa. This area, following fixation in formalin, blends into the same coloration and texture with the neighboring normal mucous membrane and is lost to the eye. This is considered to be the earliest stage of cancerous change, entirely confined to the epithelium. Eleven cases in the reported series were of this type (Fig. 6-3*B*).

Erosive Type

The lesion on the mucous membrane is slightly depressed, with an irregular edge and map-like configuration upon the mucosa. Its color is somewhat darker than the surrounding epithelium; 36 cases were of this type. In this group there were 16 carcinoma *in situ,* 17 intramucosal cancers, and 3 submucous cancers (Figs. 6-2*A*, 6-3*A*).

Plaque-like Type

The lesion is slightly elevated and appears to be a grayish-white granular plaque. Grossly, it resembles psoriatic lesions. Forty-four cases belonged to this type, among which 7 were carcinomas *in situ*, 20 invasive to the level of lamina propria, and 17 invading the submucosa (Figs. 6-1*A*, 6-3*D*).

Papillary Type

The lesions appear like raised papillomas connected to basal structures by thick pedicles. The largest papilla measures 2.5 cm across. Nine cases belonged to this category, in which one was *in situ,* four invasive to lamina propria, and four invading the submucosa (Fig. 6-3*C*).

HISTOLOGIC CLASSIFICATION OF EARLY ESOPHAGEAL CANCER

Microscopically, the early carcinoma can be classified, according to the extent of involvement, into three grades:

1. *Grade I*—Intraepithelial cancer or cancer *in situ*. The whole thickness of the epithelium is involved. Arrangement of cells in all layers is irregular with a loss of polarity. In 35 cases in the reported series, there were multiple isolated foci of carcinoma *in situ.*
2. *Grade II*—Intramucosal cancer (the earliest infiltrating cancer). In addition to intraepithelial cancer, some cells have penetrated the basement membrane into the lamina propria; 41 cases belonged to this group.
3. *Grade III.*—Submucous cancer (early infiltrating cancer). In addition to the microscopic findings described for Grades I and II, the cancerous process infiltrates downward to the submucous layer, but does not reach the muscle layer; 24 cases were of this type.

Squamous cell carcinoma can be also classified into three types according to degree of cellular differentiation and number of mitoses.

1. *Well-differentiated* (*mature type*)—The cells are mostly polygonal (prickle cell) with abundant cytoplasm and few mitoses. Well-differentiated carcinoma is usually associated with surface differ-

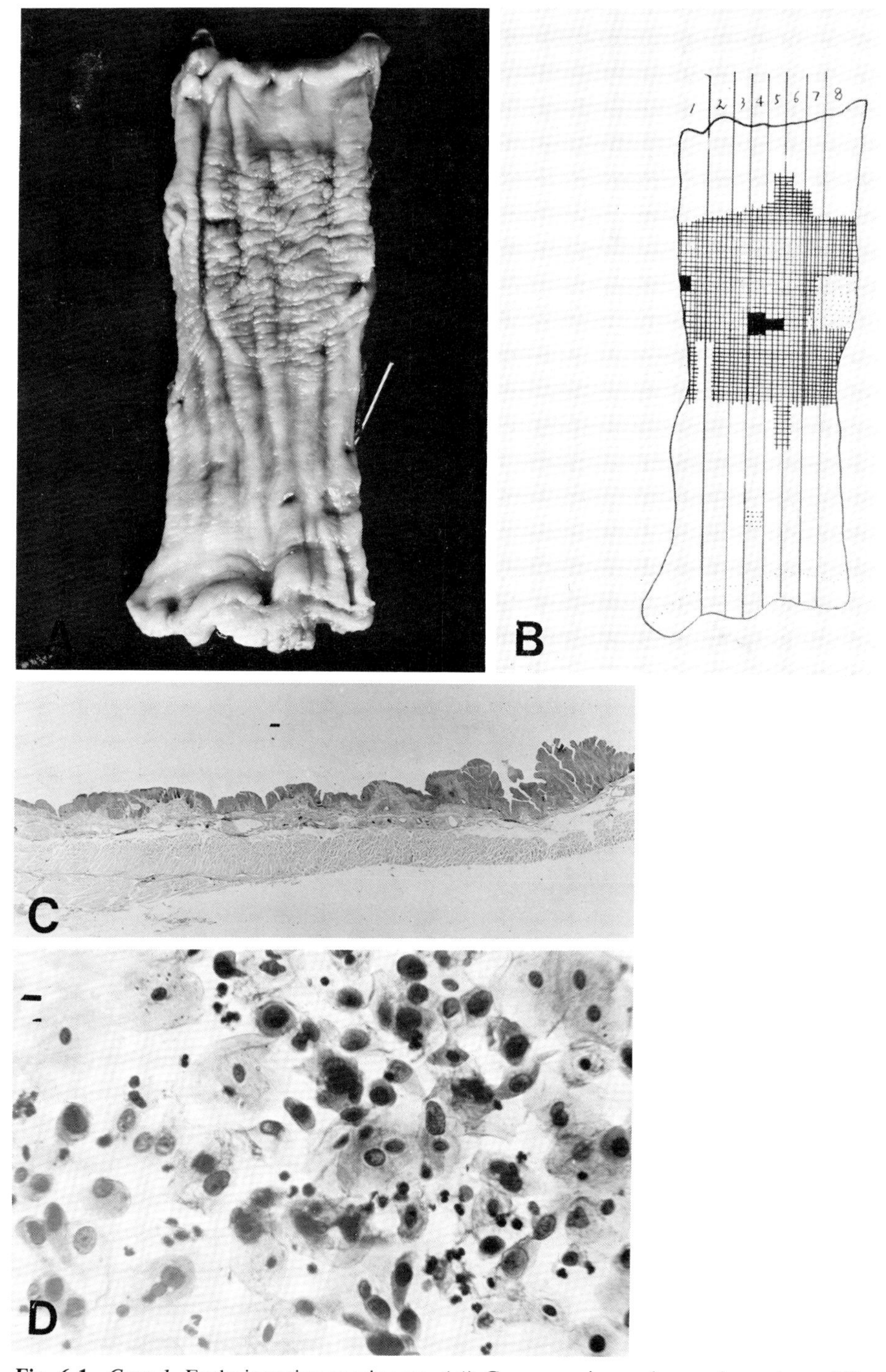

Fig. 6-1. *Case 1.* Early invasive carcinoma. (*A*) Gross specimen shows plaque type; (*B*) Diagram: early invasive carcinoma (black); carcinoma *in situ* (cross-hatched); dysplasia, (dotted). (*C*) Histologic pattern shows carcinoma *in situ* and early invasive carcinoma. (*D*) Cytologic pattern shows numerous moderately differentiated cancer cells (*D:* ×142).

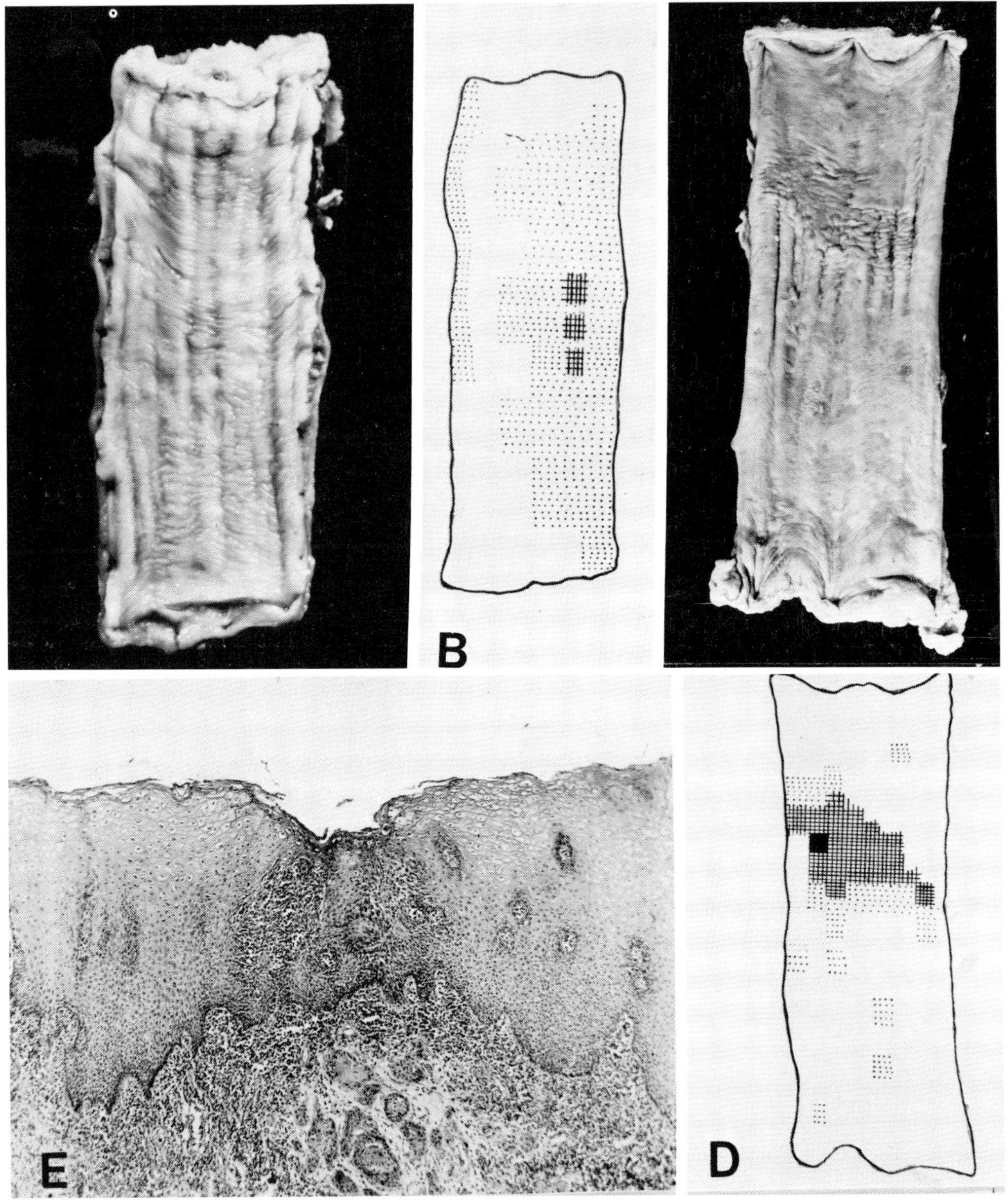

Fig. 6-2. *Case 2.* Carcinoma *in situ.* (*A*) Gross specimen shows surreptitious type. (*B*) Diagram: carcinoma *in situ* (cross-hatched); dysplasia (dotted). *Case 3.* Early invasive carcinoma. (*C*) Gross specimen shows erosion type. (*D*) Diagram: early invasive carcinoma (black); carcinoma *in situ* (cross-hatched); dysplasia (dotted). (*E*) Histologic pattern shows early invasive carcinoma (*E:* ×95).

entiation, parakeratosis, or dyskeratosis. (Fig. 6-4*A,B*).

2. *Moderately differentiated* (*intermediate type*)—The cells are smaller in size, more pleomorphic, and have less cytoplasm than the well-differentiated type. Mitoses are numerous (Fig. 6-4*C,D*).
3. *Poorly differentiated* (*immature type*)—The cells are small, sometimes spindly, with scant cytoplasm. The mitotic rate is high (Fig. 6-4*E,F*).

These three types of squamous cell carcinoma may occur simultaneously.

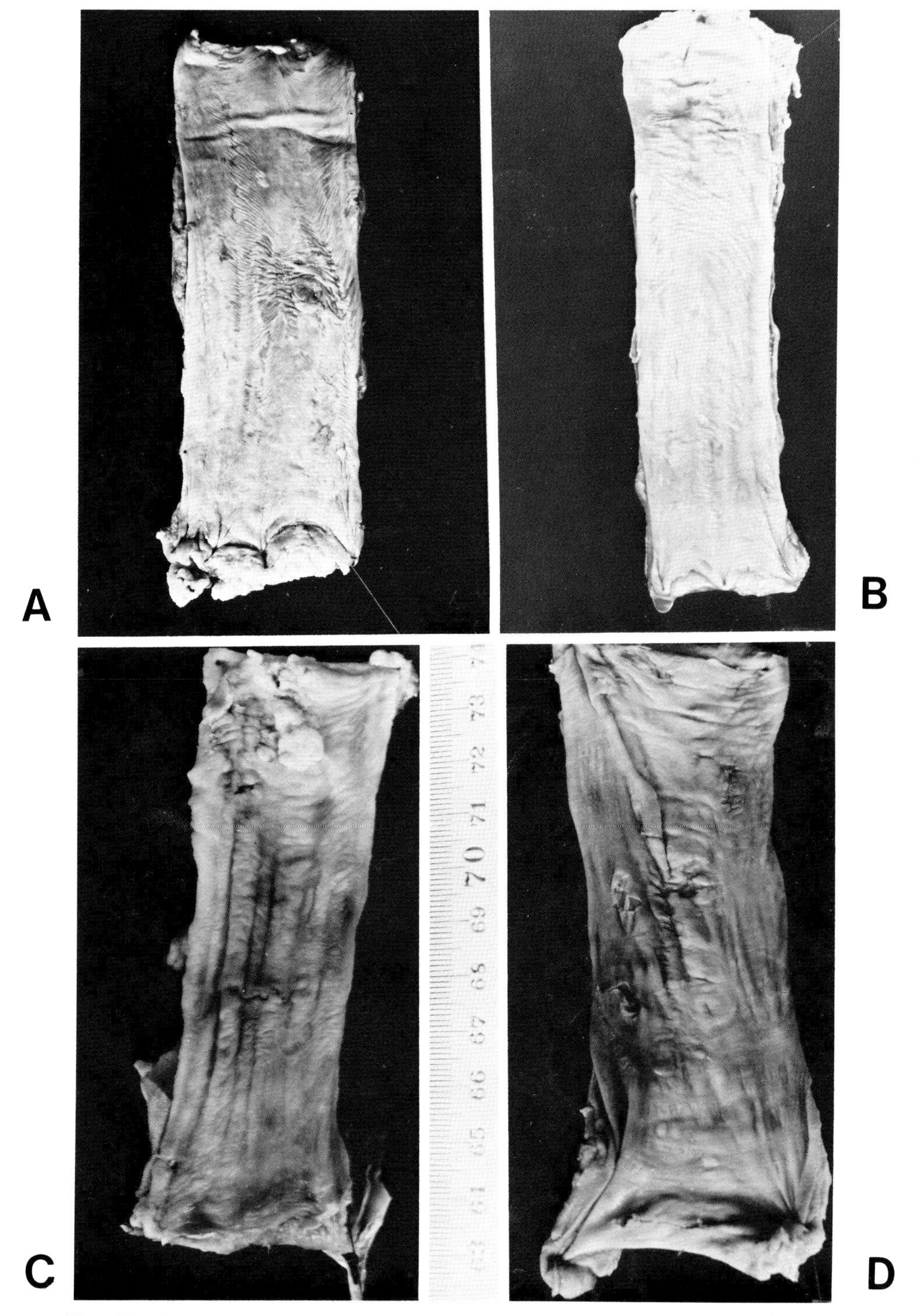

Fig. 6-3. Gross specimen shows various types of early carcinoma. (*A*) Erosion type; (*B*) surreptitious type; (*C*) papillomatous type; (*D*) plaque type with a small ulcer.

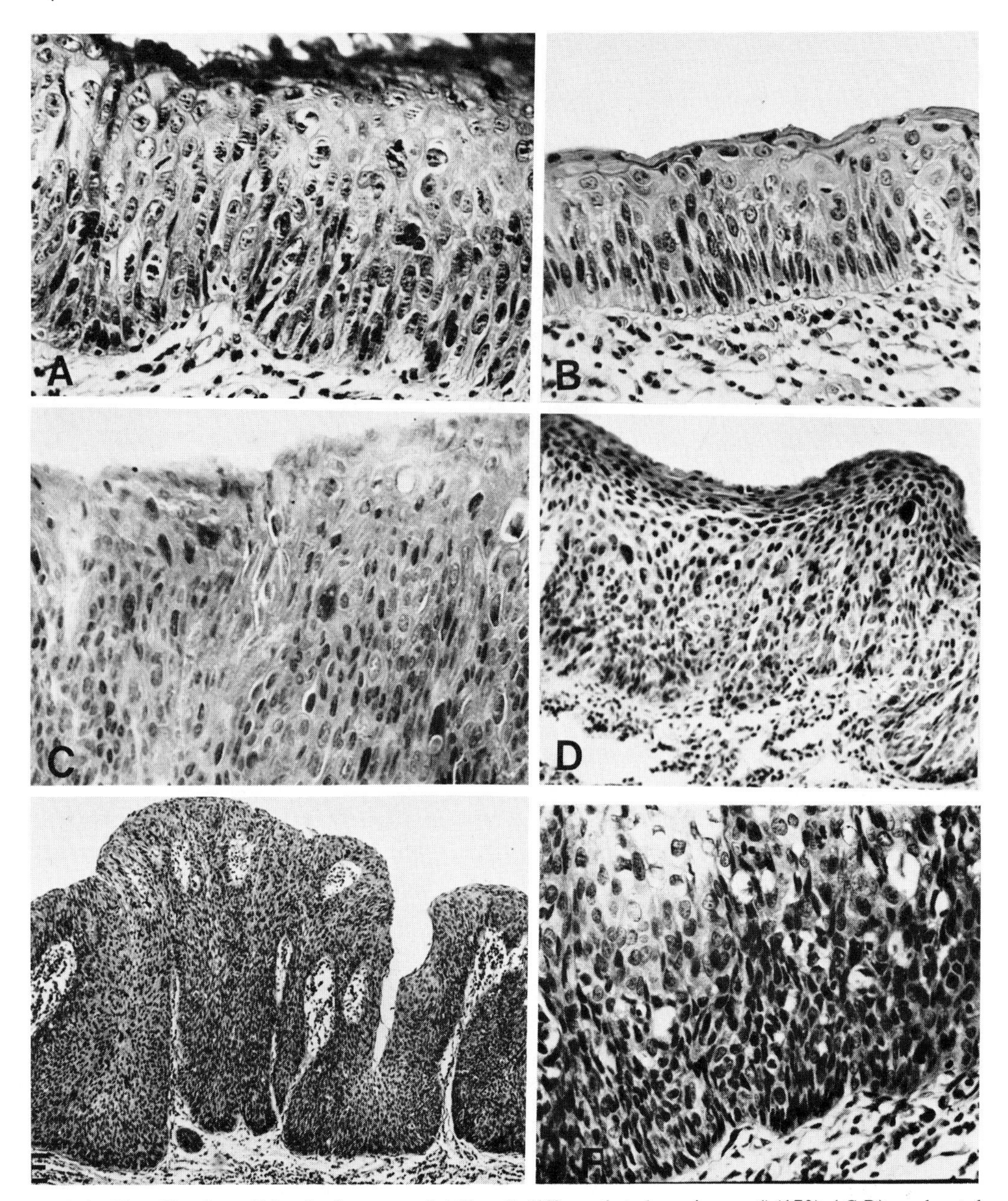

Fig. 6-4. Classification of histologic pattern (*A,B*) well-differentiated carcinoma (×170); (*C,D*) moderately differentiated carcinoma (*C:* ×170; *D:* ×85); (*E,F*) poorly differentiated carcinoma (*E:* ×45; *F:* ×170).

CYTOPATHOLOGIC FEATURES OF EARLY ESOPHAGEAL CANCER

Shu[30] investigated the cytomorphologic features in 100 cases of early esophageal cancer. Three smears per case were evaluated.

Smear Background

Of 100 cases of early esophageal cancer, 72 showed mild inflammation, 23 showed moderate inflammation, and five showed severe inflammation. Although all of the smears had some degree of inflammation, the extent was generally less than in advanced cancer. There was also less necrosis noted. Usually, the background of the smear was clear with few inflammatory cells and erythrocytes. Fungi were frequently present.

Dysplastic Cells

Ninety-six of the cases had numerous dysplastic cells in the background, usually showing the pattern of severe dysplasia. In four cases there were few dysplastic cells.

Cancer Cells

The number of obvious cancer cells per case may be very small, with less than 10 cells in 36 cases. A single cancer cell was observed in three of these cases. The findings, summarized in Table 6-1, emphasize the need for a very careful screening of esophageal smears.

Arrangement and Configuration of Cancer Cells

In 84 cases the cancer cells appeared as single cells only. In 16 cases cell aggregates were also observed. Using the cytologic criteria outlined above, moderately-differentiated cells were observed in 53 cases, poorly differentiated cancer cells in 38 cases, and well-differentiated cells in 6 cases.

Nuclear Morphology and Morphometry in Esophageal Cancer

Lee[36] studied the features of 1100 cancer cells from smears of 29 patients with early esophageal carcinoma. Cell measurements were made with 95X (oil immersion) objective and a 5X ocular with a built in micrometer.

Number of Nuclei: Cancer cells with a single nucleus predominated (91%), binucleated cancer cells were fairly frequent (7.3%), and multinucleated cells were rare (1.7%).

TABLE 6-1
Number of Cancer Cells in 100 Cases of Early Esophageal Cancer

No. of cancer cells in 3 smears/case	No. of cases
Less than 10	36
11–25	29
26–49	16
Over 50	19

Nuclear Chromatin Pattern

Lee[36] separated the nuclear chromatin into three fundamental patterns: coarse granular, thick reticular, and clumped. The results were as follows: clumped was observed in 10.5% of the cells; coarse granular in 9.5%; thick reticular in 43.2%; clumped and coarse granular in 4.3%; coarse granular and thick reticular in 12.1%; clumped and thick reticular in 20.5%. As can be seen, the most common chromatin pattern is thick reticular, which is frequently observed in early esophageal cancer cells.

Nucleoli

The number of nucleoli varied from 1 to 6 per cancer cell. The number of nucleoli also varied with the developmental stage of the disease, multiple nucleoli being more common in cells from advanced cancer. In general, the malignant cells showed prominent and multiple nucleoli. In the present study 31.72% of the cancer cells from early cancer had indistinct nucleoli.

MORPHOLOGIC CHARACTERISTICS OF EARLY SQUAMOUS CELL CARCINOMA OF THE ESOPHAGUS

The cytologic picture of early esophageal cancer can be summarized as follows.[12] (See Figs. 6-5 to 6-16.)

1. There is a preponderance of moderately differentiated cancer cells. The

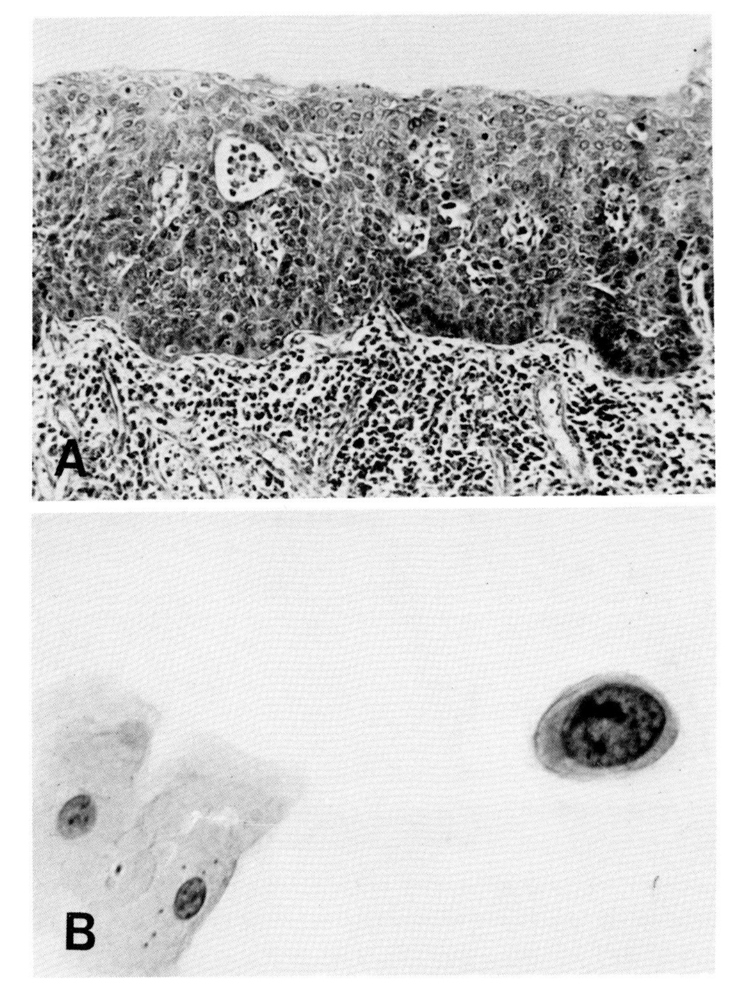

Fig. 6-5. *Case 4.* The first case of esophageal cancer detected by using balloon technique. (*A*) Histologic pattern shows carcinoma *in situ* (×200); (*B*) single cancer cell (×400).

rounded types are the most frequently observed.

2. The nuclear border is even and smooth, and aberrant nuclei are rarely seen.
3. There are only a limited number of cancer cells scattered about in the smear.
4. In more than 96% of the smears, there exists a great number of cells with severe dysplasia, most of which belongs to the basal dyskaryotic type. The basal, intermediate, and superficial dyskaryotic cells show irregular nuclei. The fiber type dyskaryotic cells and the premature cells with aberrant nuclei are more abundant than those observed in marked hyperplasia.
5. The background of the smear is usually clean with scanty inflammatory cells, degenerated cells, and red blood cells.

AGE AT DETECTION[38]

In cases of early esophageal carcinoma, based on cellular evidence, the mean age at detection of 93 consecutive cases was 56.30 years (57.04 in male and 55.67 in female). The distribution in various age groups is shown in Figure 6-17.

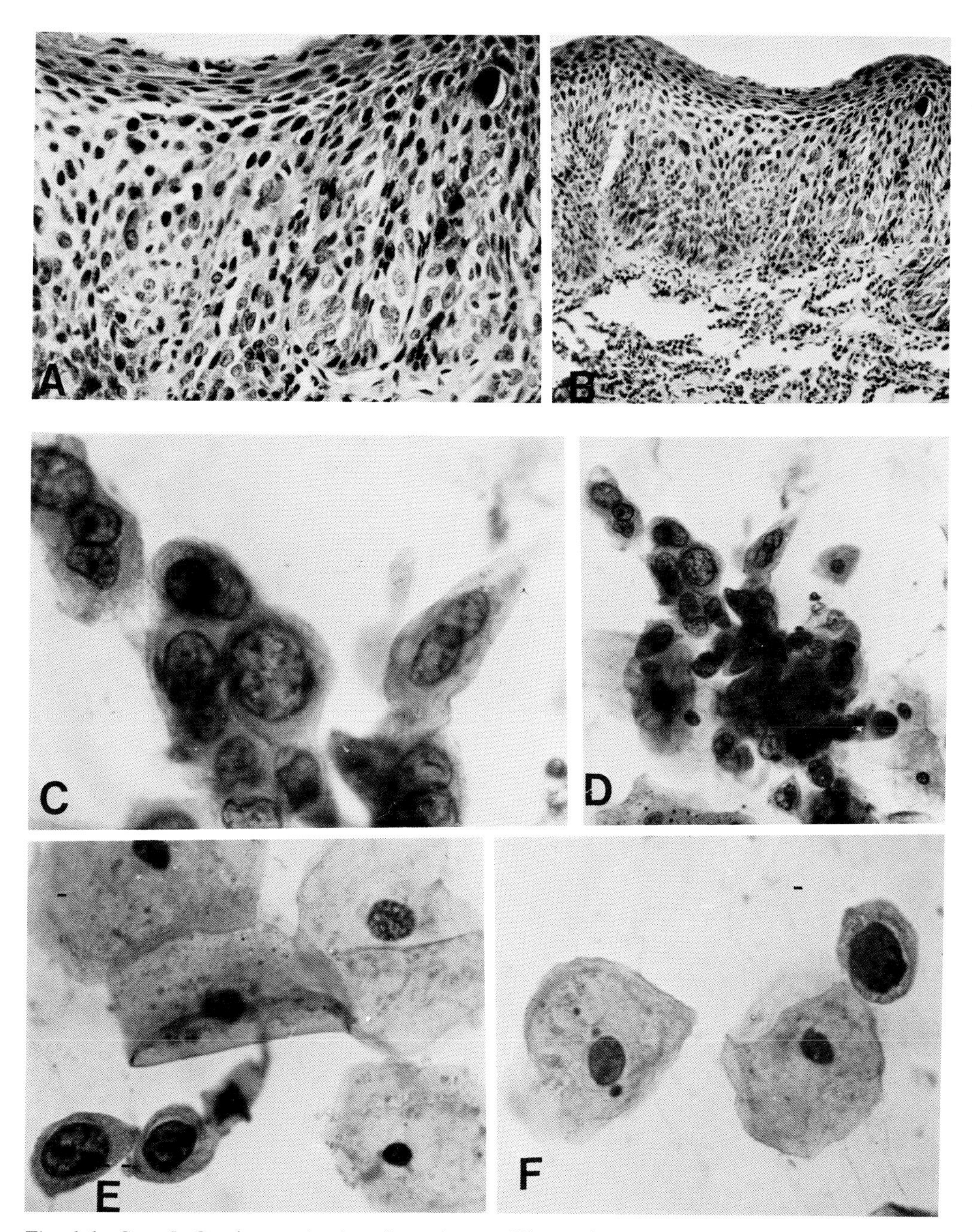

Fig. 6-6. *Case 5.* Carcinoma *in situ* of esophagus. Histologic pattern (*A:* ×180; *B:* ×90). Cytologic pattern shows various types of moderately differentiated cancer cells (*C,E,F:* ×380; *D:* ×180).

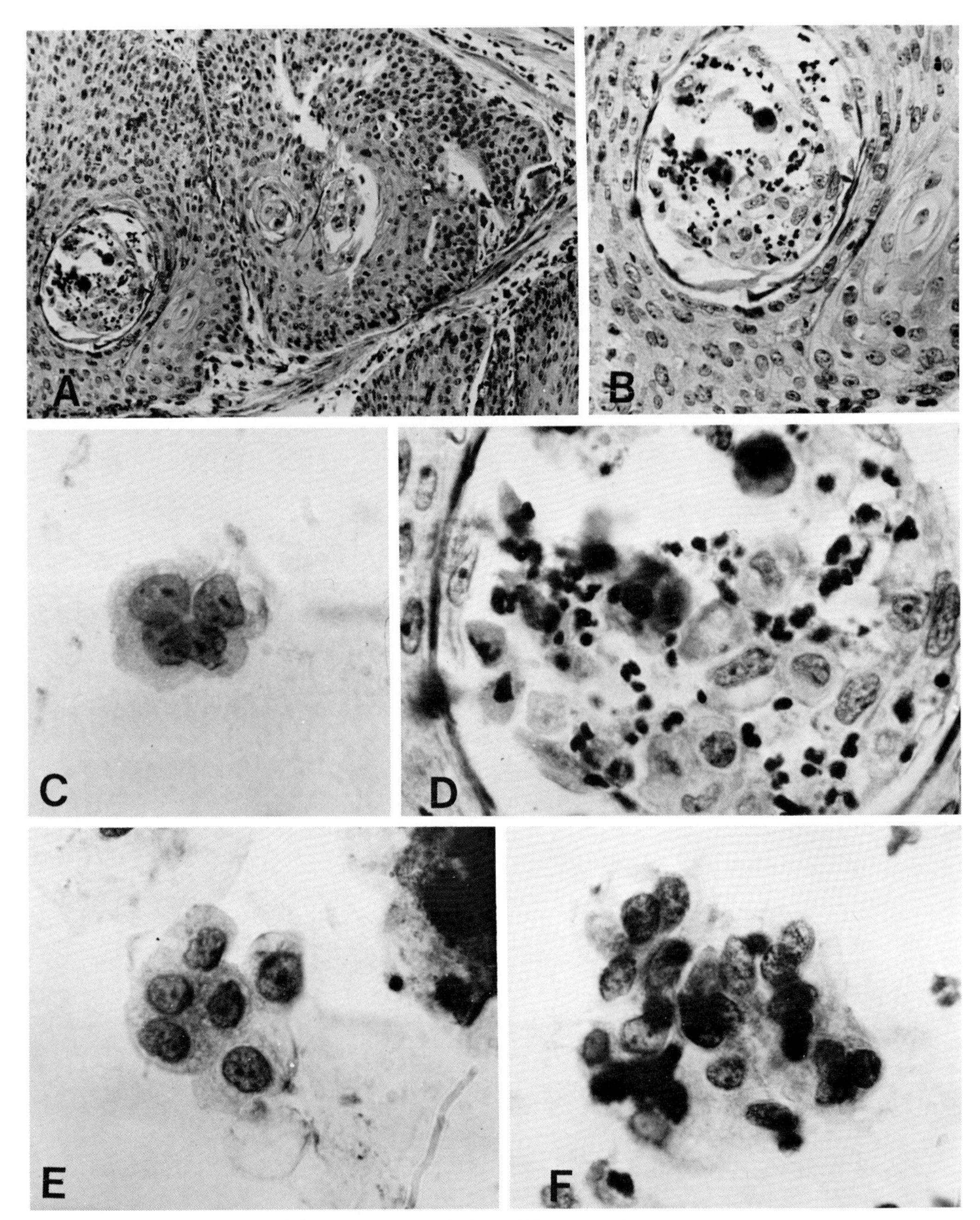

Fig. 6-7. *Case 6.* Early invasive carcinoma. (*A*) Well-differentiated squamous cell carcinoma of esophagus. Isolated keratinizing pearl is present (×90). (*B*) Cancer cells are seen in the central space of a cancerous nest (×180); (*D*) same as above (×360). (*C,E,F*) Moderately differentiated cancer cells (×360).

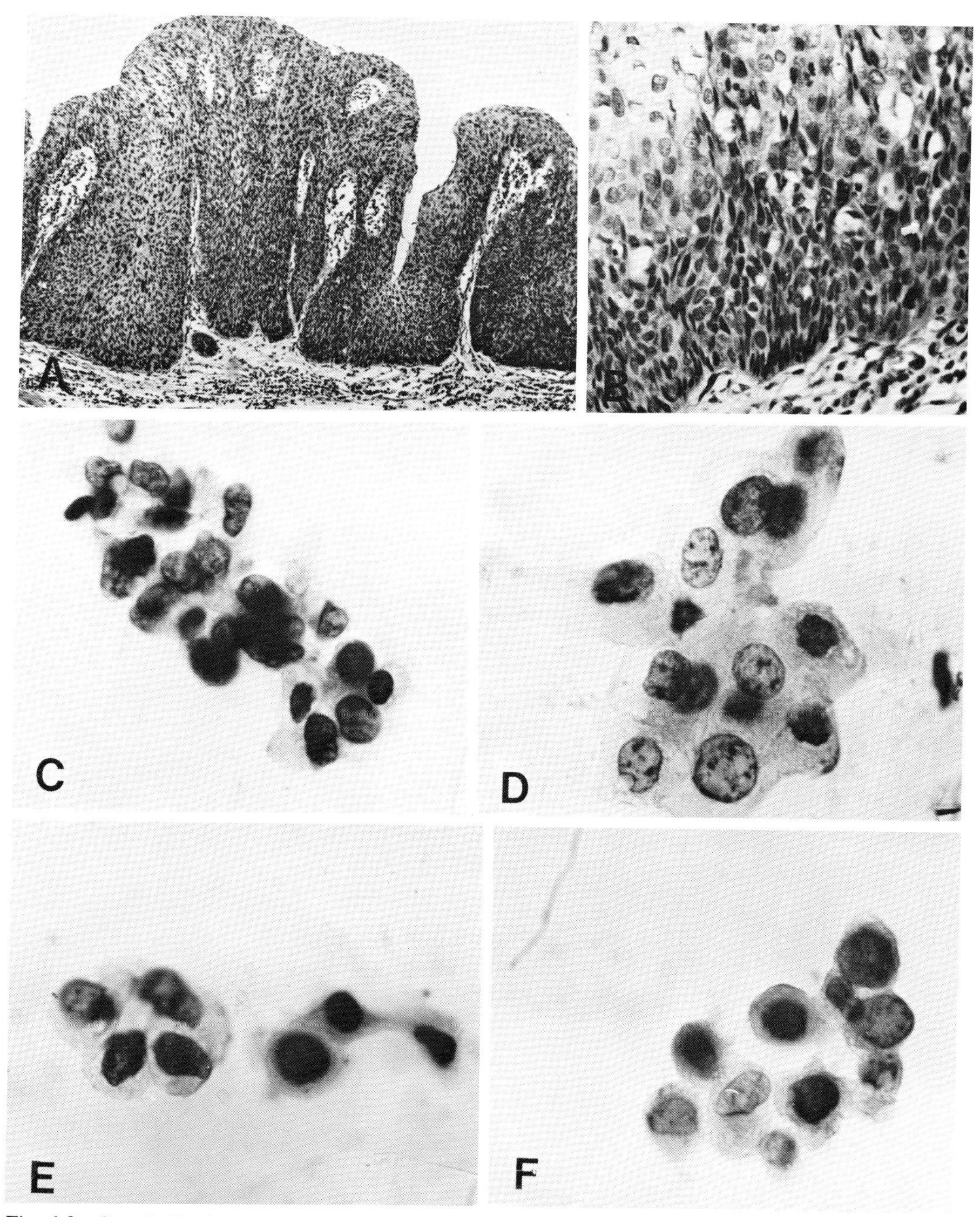

Fig. 6-8. *Case 7.* Carcinoma *in situ* of esophagus. Histologic pattern (*A:* ×90, *B:* ×180); cytologic pattern (*C–F:* ×360).

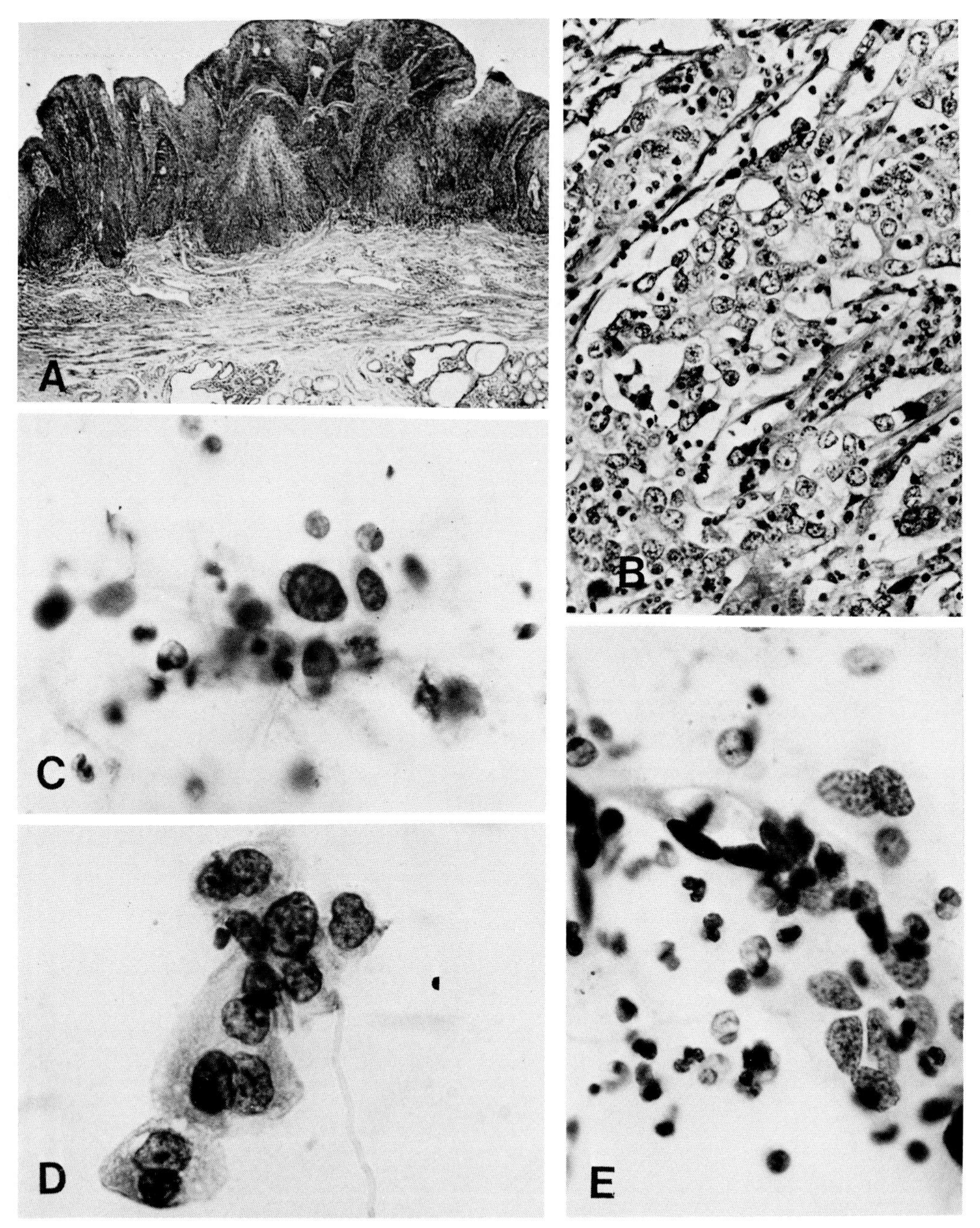

Fig. 6-9. *Case 8.* Early invasive carcinoma of esophagus, squamous cell carcinoma (*A*) poorly differentiated squamous cell carcinoma (×45); (*B*) gland-like structure (×180); (*C,D,E*) moderate and poorly differentiated cancer cells. Note the spindle shapes (×360).

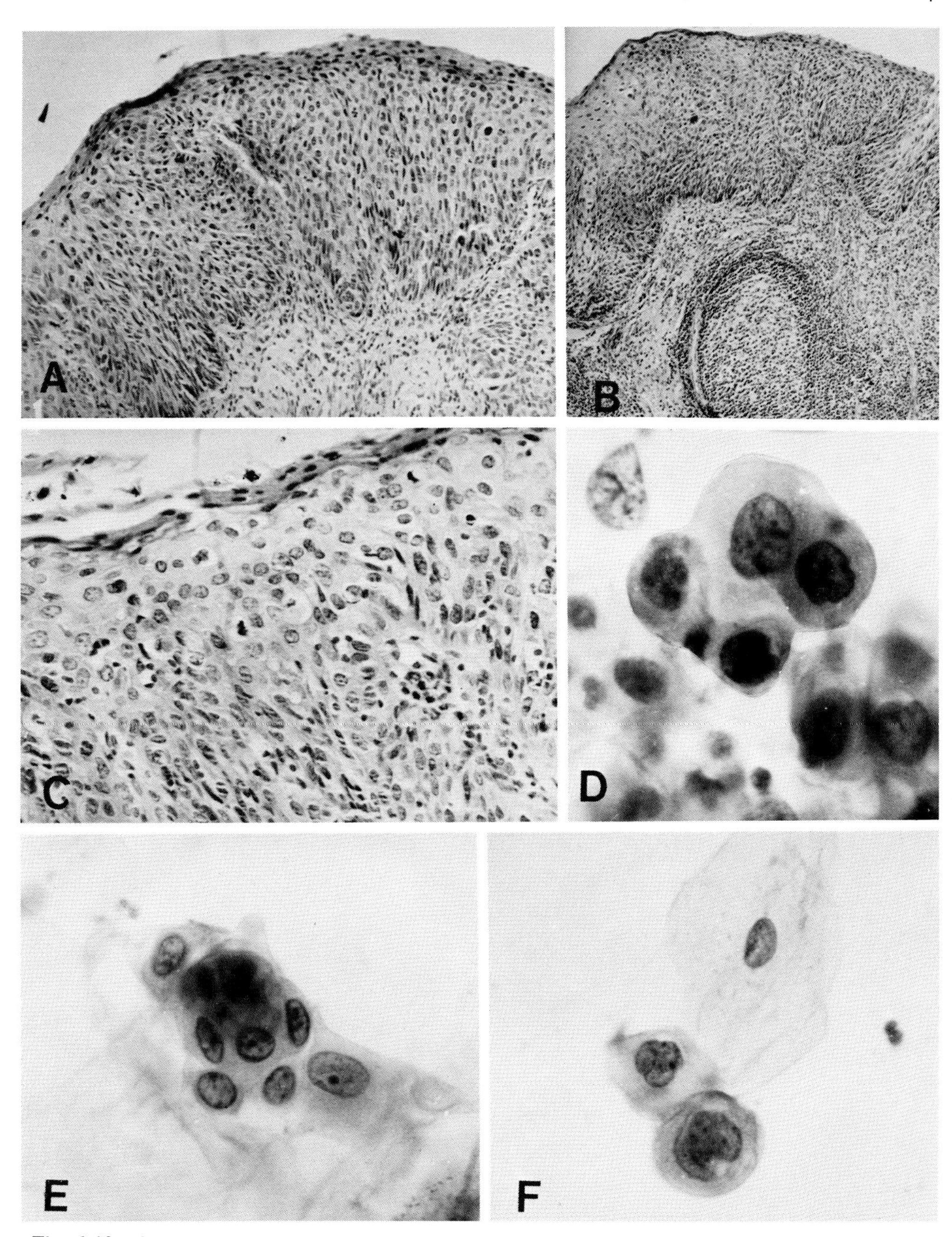

Fig. 6-10. *Case 9.* Carcinoma *in situ* of esophagus. (*A*) Well-differentiated squamous cell carcinoma (×90). (*B*) Same as *A;* note prominence of lymphocytic follicular reaction (×45). (*C*) Surface reaction overlying keratinizing squamous cell carcinoma (×180). (*D,E,F*) Round to oval, moderately differentiated cancer cell (×360).

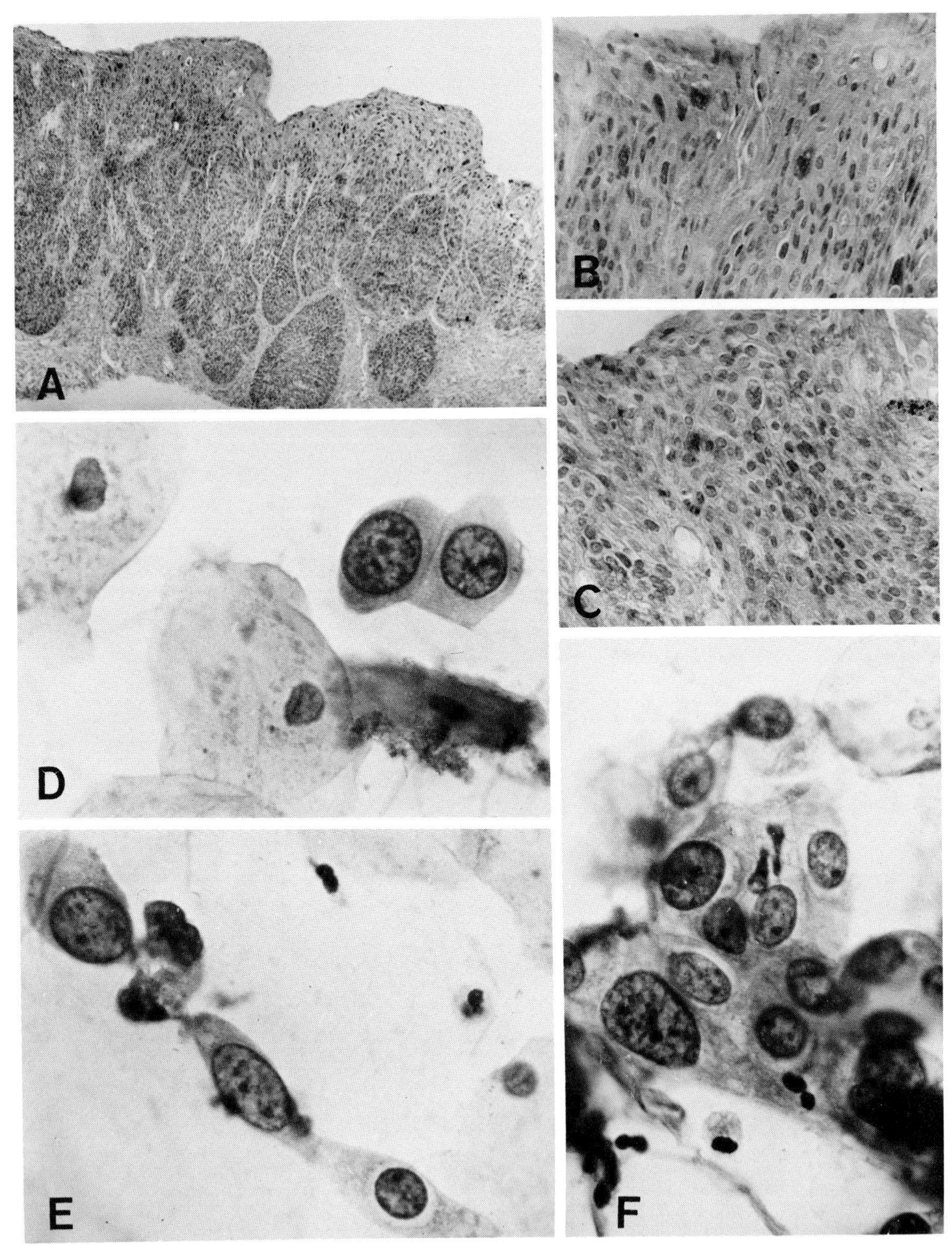

Fig. 6-11. *Case 10.* Early invasive carcinoma. (*A*) Well-differentiated squamous cell carcinoma. Note invasive nest of carcinoma (×45). (*B,C*) Surface reaction overlying keratinizing squamous cell carcinoma (×90). (*D,E*) Well-differentiated and moderately differentiated cancer cells (×360).

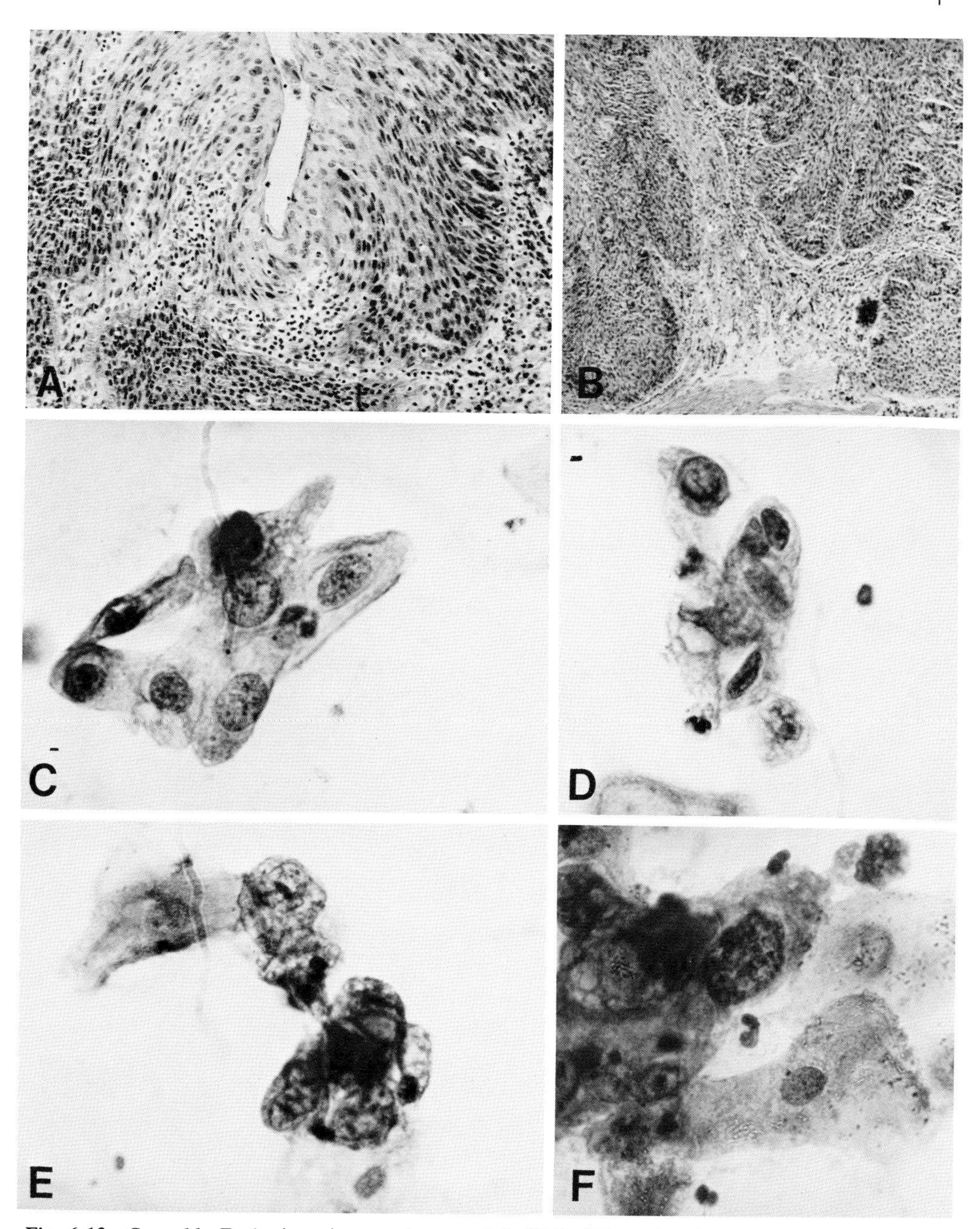

Fig. 6-12. *Case 11.* Early invasive carcinoma. (*A*) Well-differentiated squamous cell carcinoma (×90). (*B*) Carcinomatous tissue involves submucosa to muscle layer (×45). (*C,D,F*) Well- and moderately differentiated cancer cells (×360). (*E*) Poorly differentiated cancer cells (×360).

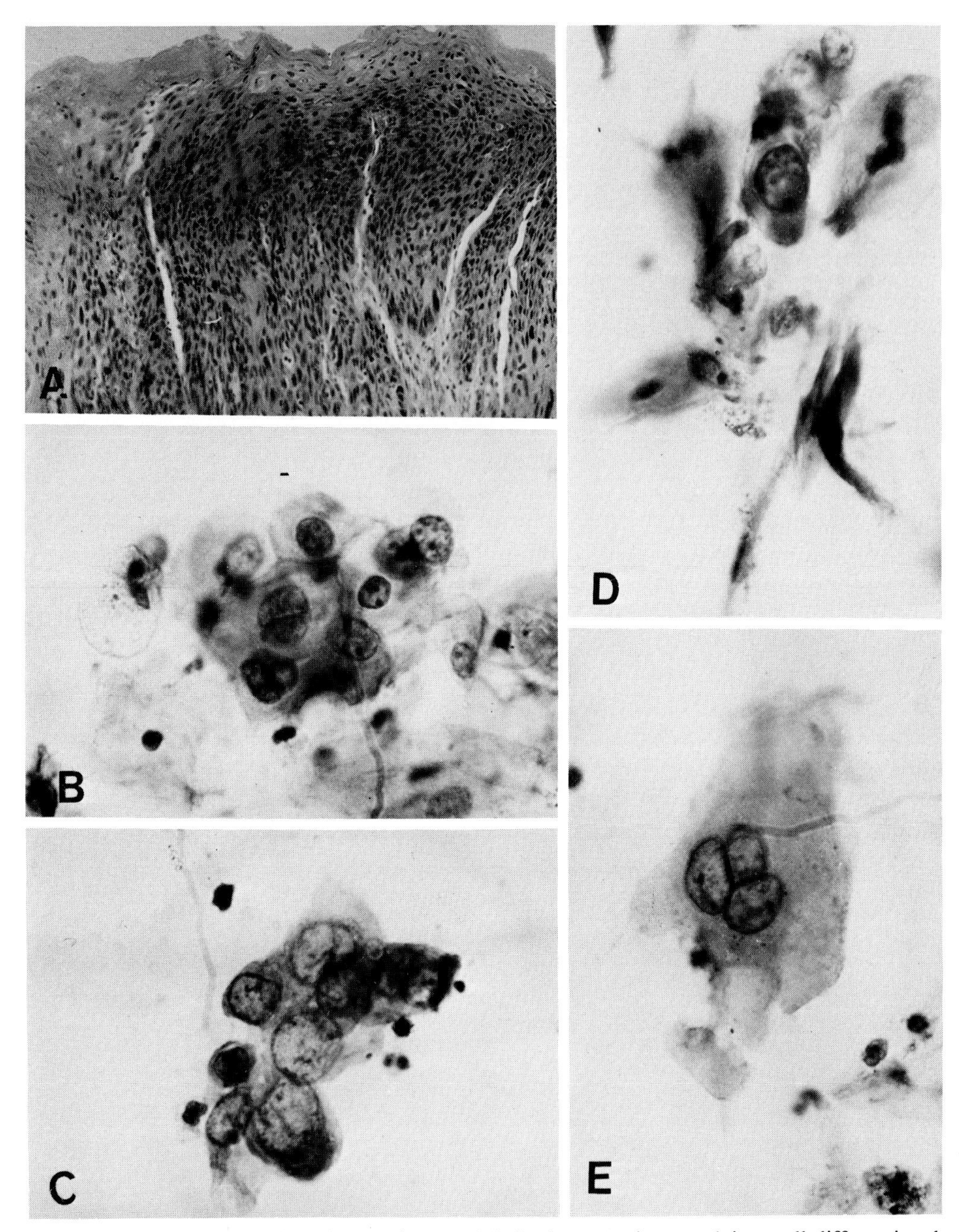

Fig. 6-13. *Case 12.* Early invasive carcinoma. (*A*) Surface reaction overlying well-differentiated squamous cell carcinoma (×90). (*B*) Moderately differentiated cancer cells (×360). (*C*) Poorly differentiated cancer cells (×360). (*D*) Well-differentiated cancer cells (×360). (*E*) Multinucleated giant cell of esophageal epithelium (×360).

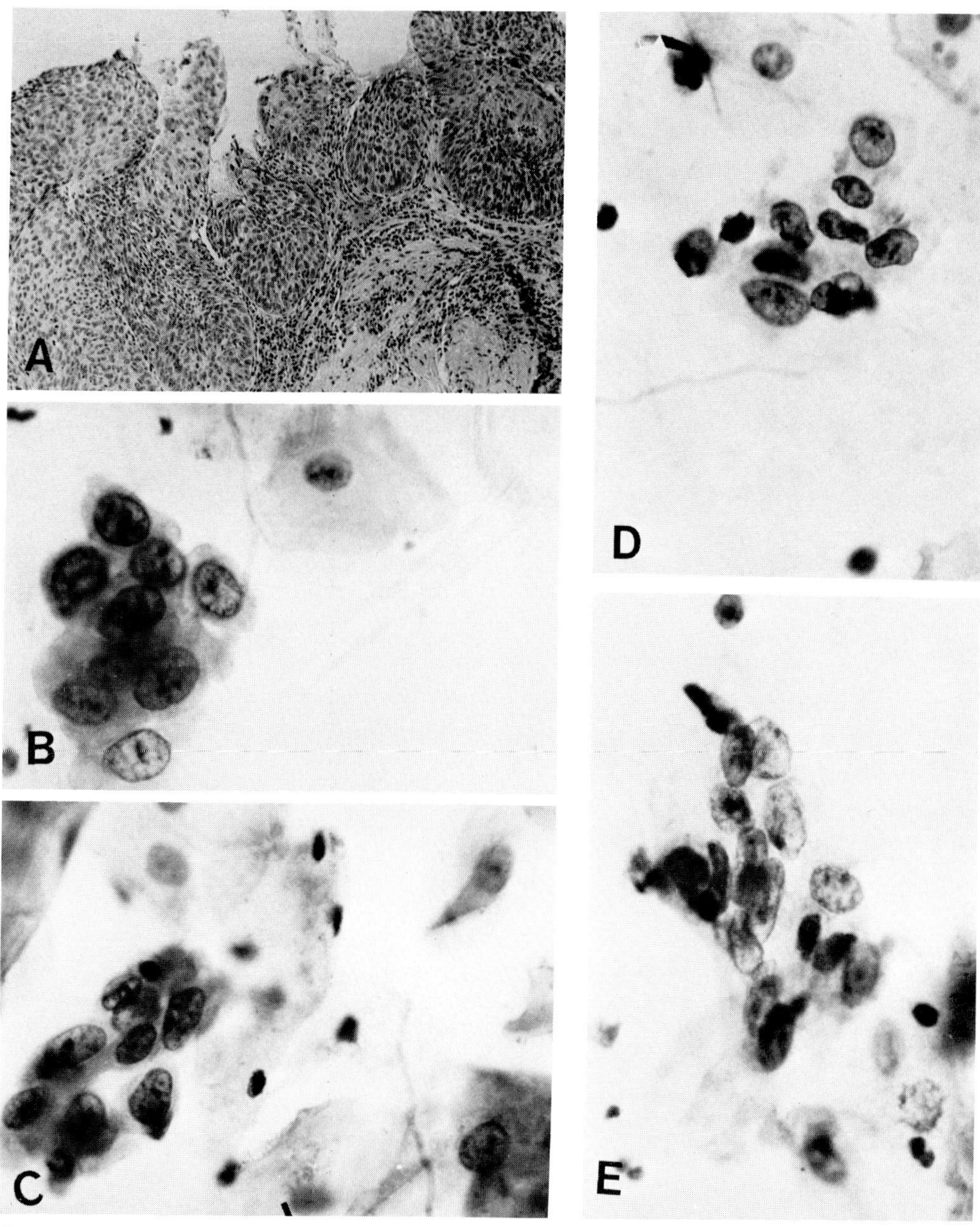

Fig. 6-14. *Case 13.* Early invasive carcinoma. Histologic pattern (*A:* ×90). Cytologic pattern (*B–E:* ×360).

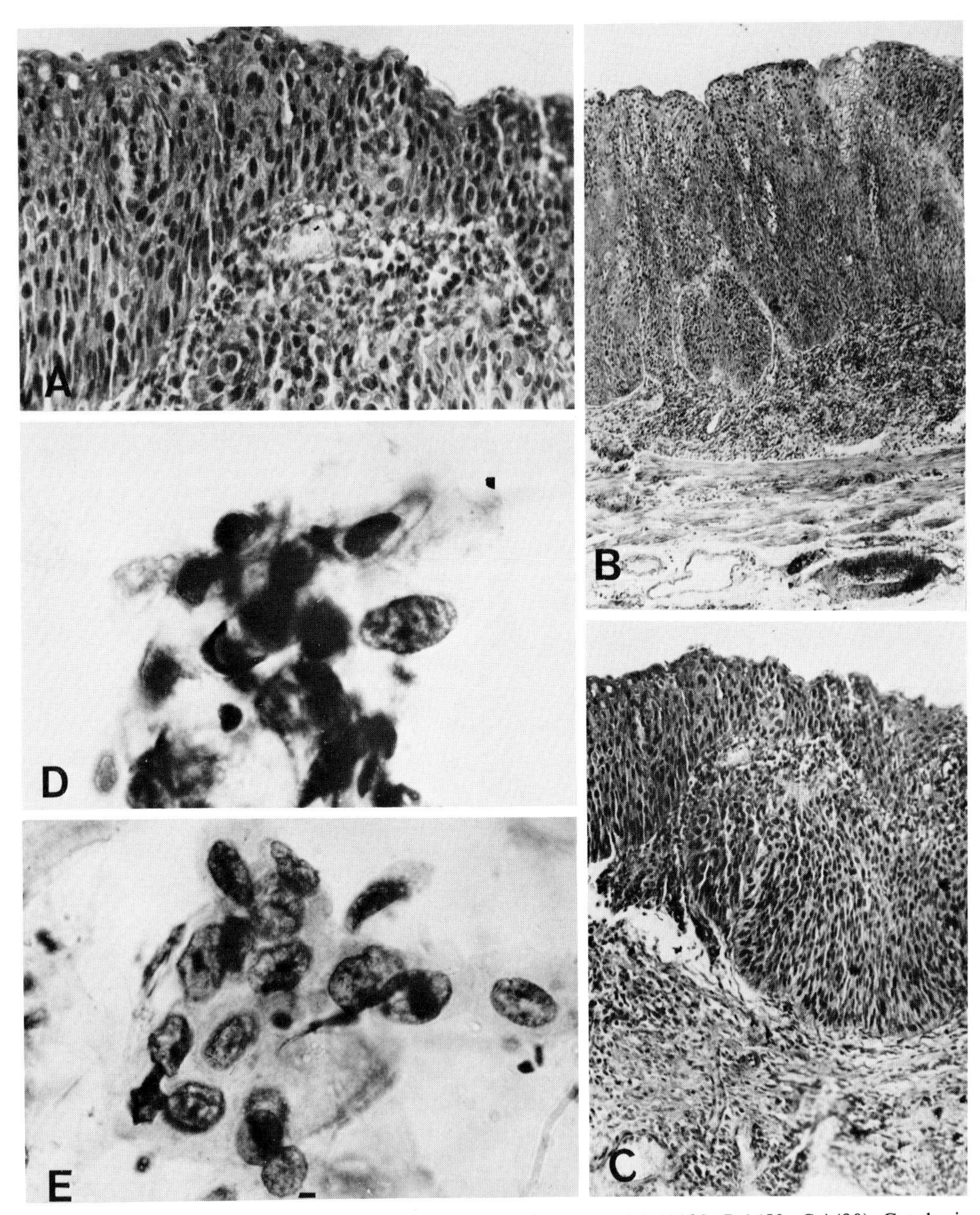

Fig. 6-15. *Case 14.* Early invasive carcinoma. Histologic pattern (*A:* ×180; *B:* ×50; *C:* ×90). Cytologic pattern shows poorly differentiated cancer cells (*D,E:* ×360).

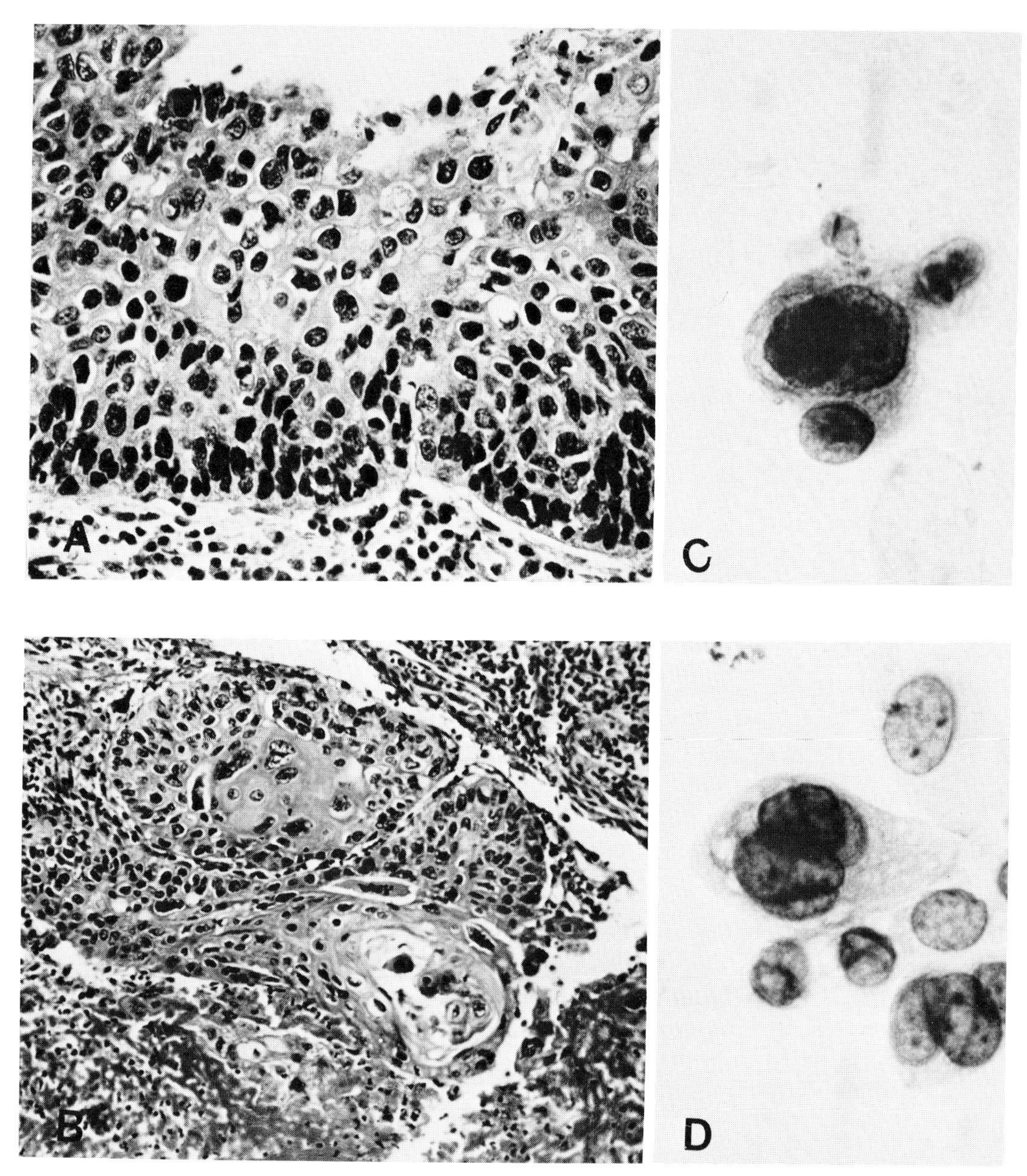

Fig. 6-16. *Case 15.* Early invasive carcinoma. Histologic pattern (*A:* ×270; *B:* ×135). Cytologic pattern (*C,D:* ×540).

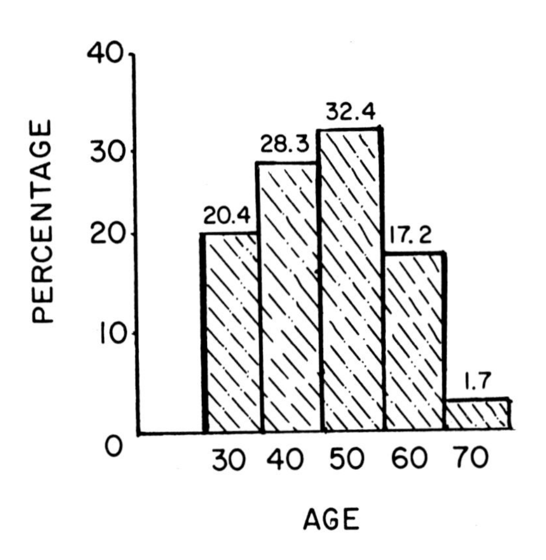

Fig. 6-17. Distribution of various age groups in early cancer of the esophagus.

The mean age at detection of 41 consecutive cases of moderately advanced cancer was 60.30 years. The mean age at death of esophageal carcinoma was 63.49 years. The mean age at detection of severe dysplasia was 51.70 years. Chronologic progression of the lesion can be postulated by this epidemiologic evidence:

51.70 severe dysplasia	—	56.30 early carcinoma	—	60.30 moderately advanced carcinoma	—	63.49 death age

SYMPTOMS OF EARLY CANCER

The symptoms of early cancer in 170 cases of early esophageal cancer were as follows: 50.6% had difficulty swallowing; 48% had mild pain when swallowing; 18.2% had discomfort behind the sternum; 15.3% had a feeling of a "foreign-body" when swallowing; 11.7% had epigastralgia; 8.8% felt a constricting of the throat; 3.5% belched on food intake; and 7.6% were asymptomatic.[37]

ENDOSCOPIC EXAMINATION

An endoscopic study of 128 cases of early esophageal cancer had the findings in Table 6-2.

TABLE 6-2
Endoscopic Study of Early Esophageal Cancer

Lesion	No. of cases	%
Local erosion	53	41.42
Local plaque	17	13.30
Local hyperanemia	30	23.36
Small nodular or glandular	11	8.61
Surface ulcer	3	2.36
Polypoid change	1	0.80
Uncertain lesion	13	10.15
Total	128	100.00

CHAPTER 7

Advanced Cancer of the Esophagus

GROSS SPECIMEN

A study of 100 cases[39] suggested that esophageal cancer could be classified by gross examination into four different types. These were medullary (Fig. 7-1*D*), fungating (Fig. 7-1*C*), ulcerating (Fig. 7-1*A*), and sclerrous (later changed to stenosing) (Fig. 7-1*B*). This method of classification has since been adopted extensively in China.

Of 200 cases reported by Wu,[40] 117 cases (58.5%) were classified as medullary type, 34 cases (17%) as fungating, 22 cases (11%) as ulcerating, and 17 cases (8.5%) as stenosing. In addition, 10 cases (5%) were listed as unclassifiable.

HISTOLOGY

In China, esophageal carcinoma has been categorized into mature, intermediate, and immature types (comparable to well, moderately, and poorly differentiated).

In the mature type, the cancer cells are polygonal. The neoplasms contain epithelial pearls, keratinization, and some intercellular bridges. Mitoses are scanty and cytoplasm is abundant (Fig. 7-2*A*).

In the immature type, the esophageal wall is invaded by cancer infiltrating between muscle fibers. Epithelial pearls, keratinization, and intercellular bridges are not present. Mitoses are common. The cells are spindled, small, or pleomorphic. There is a marked variation in size and shape. The cytoplasm of the cancer cell is scanty (Fig. 7-2*B*).

The histologic features of the intermediate type are more advanced than the immature types, but less advanced than the mature types.

CYTOLOGY

Cellular Morphology

There is a large number of moderately differentiated cancer cells; however, keratinizing and bizarre cancer cells are also commonly observed (Fig. 7-3*C,D*). Variability in size, shape, and color of cancer cells is more pronounced in advanced lesions (Fig. 7-3*B,C,D,* 7-4*C,D*).

Nucleus of the Cancer Cell

Most nuclei are round or oval shaped. The incidence of lobulated and multinucleated cell forms is much higher in advanced cancer (Fig. 7-4*A*), as is the percentage of clumped, coarse, granular, and mixed types.

Arrangement of Cancer Cells

Cancer cells have been observed both in isolation and in clumps (Fig. 7-3*B*). Compared to smears of early cancer, smears from advanced and late cancer had relatively fewer isolated cells than clumped cells.

Number of Cancer Cells

If the cytologic material is obtained properly by the abrasive balloon, then a large number of cancer cells will be visible in the smear.

The Smear Background

A "dirty" background is characteristic of advanced esophageal cancer. The presence of granular and proteinaceous material originates from tissue destruction and necrosis. Fragmented erythrocytes and/or inflammatory cells also can be present. This background is often associated with fungal forms and/or plant cells.

SUMMARY OF DIAGNOSTIC CYTOLOGIC FEATURES

TABLE 7-1
Comparative Cytologic Features of Early and Advanced Cancer of the Esophagus

	Early	Advanced
Cancer cell		
Arrangement	Mainly single isolated	Single isolated in groups or clusters
Pleomorphism	Less common	Common
Number	Usually less than 50	Usually greater than 100
Keratinizing	Less common	Common
Nucleus		
Bizarre	Less common	Common
Lobulated or multinucleated	Uncommon	More often
Chromatin pattern	Mainly thick, reticular	Predominantly coarsely granular, many opaque forms (many opaquely hyperchromatic forms)
Macronucleoli	Uncommon	More common
Multinucleoli	Uncommon	More common
Dysplastic cells	Many	Few
Tumor diathesis background	Uncommon	Common

TABLE 7-2
Comparative Cytologic Features of Severe Dysplasia and Early Cancer of the Esophagus

	Severe dysplasia	Early cancer
Nuclear–cytoplasmic ratio	Slightly increased	Markedly increased
Nuclear enlargement	Slight	Marked
Nuclear pleomorphism	Rare	Few
Thickness of nuclear membrane	Slightly thickened	Markedly thickened (0.75 μ), regular to irregular
Chromatin pattern	Finely granular to granular, even and regular	Thick reticular or coarse granular uneven (irregular)
Cancer cell	Suspicious only	Present
Severe dysplastic cells	Many	Few to many
superficial layer	100% of cases	96% of cases
intermediate layer	92% of cases	78% of cases
basal layer	34% of cases	100% of cases
Tumor diathesis	Absent	Uncommon

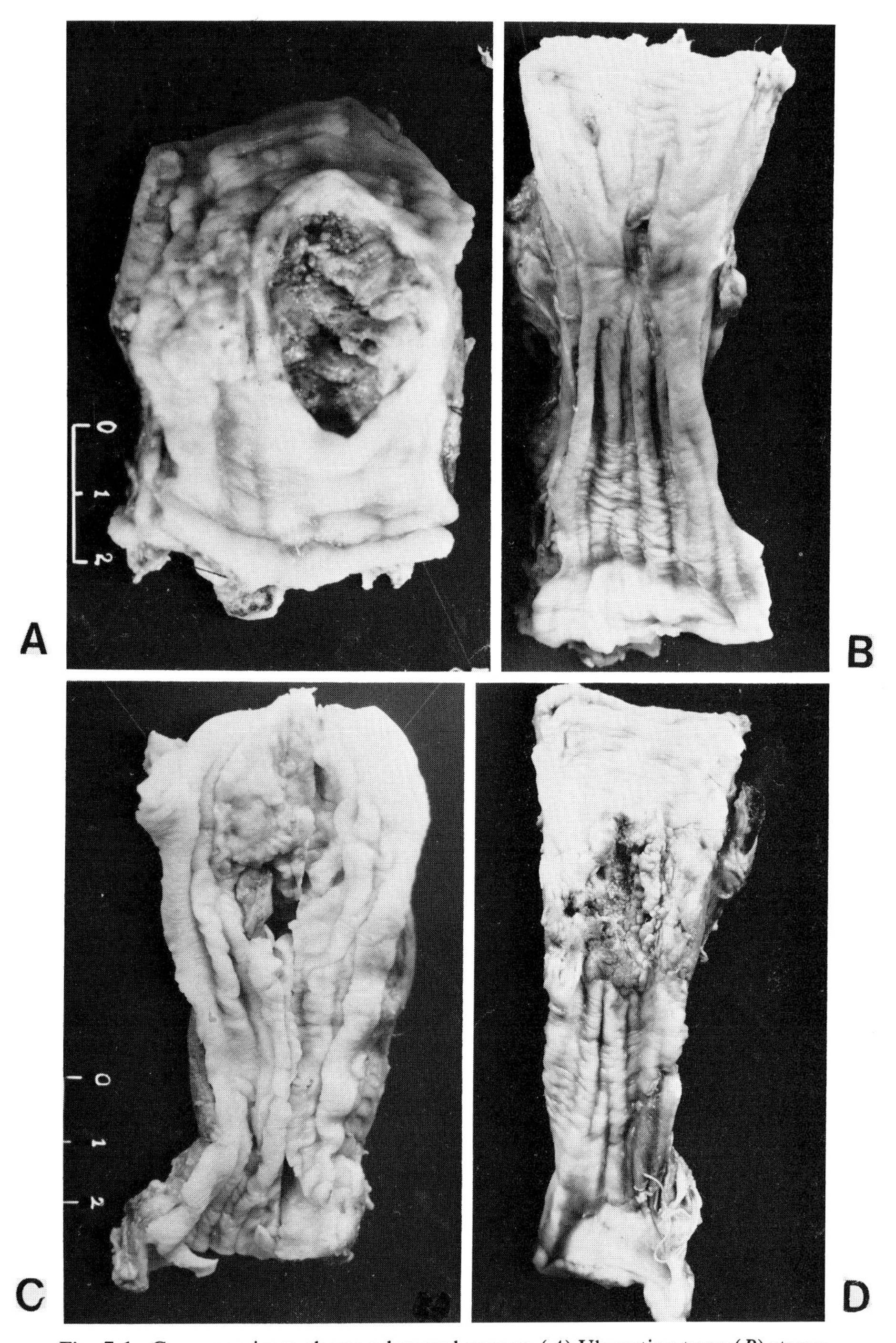

Fig. 7-1. Gross specimen shows advanced cancer. (*A*) Ulcerating type; (*B*) stenosing type; (*C*) fungating type; (*D*) medullary type.

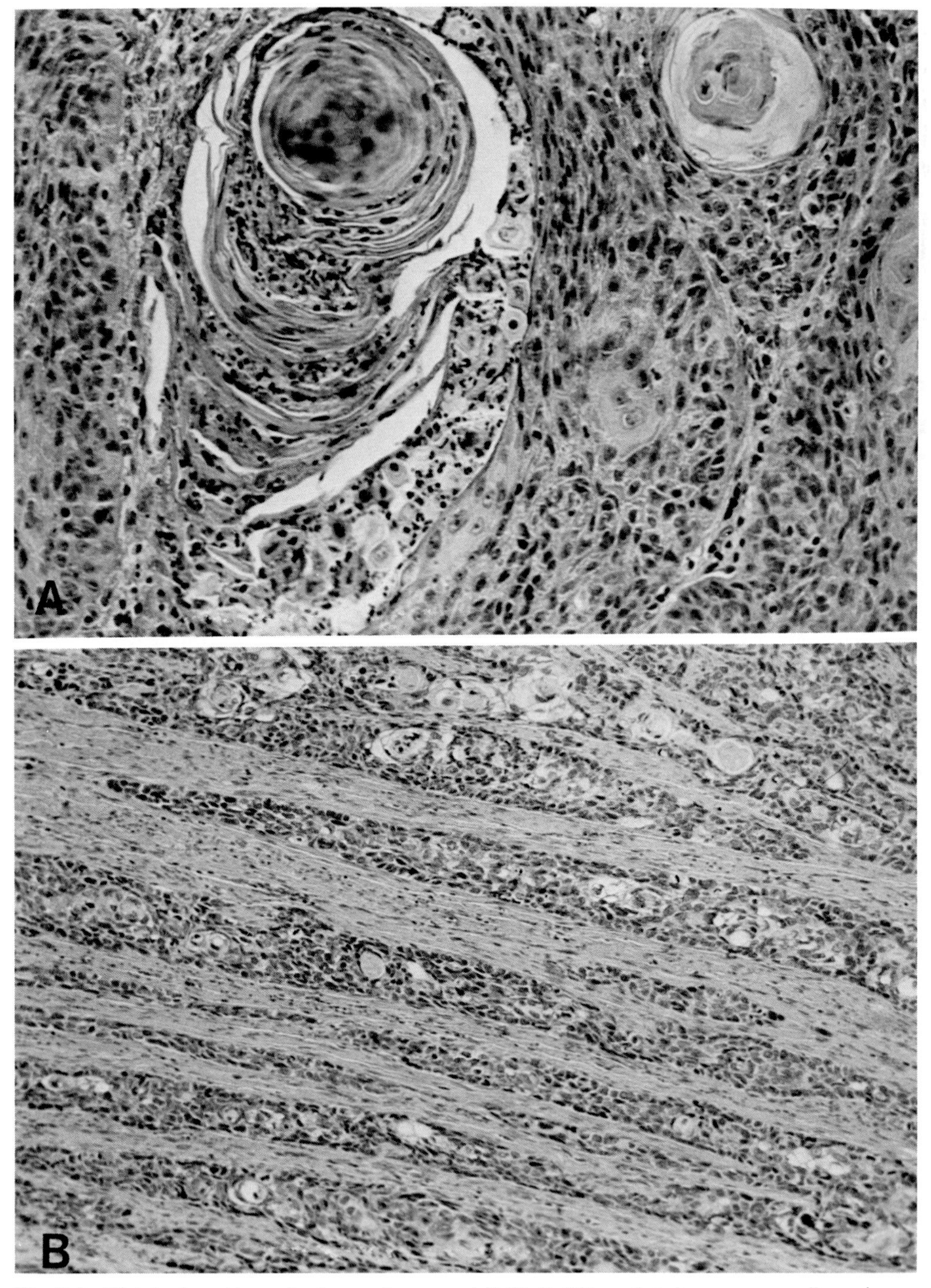

Fig. 7-2. Histologic pattern of advanced cancer. (*A*) Well-differentiated squamous cell carcinoma with keratinizing pearl formation. (*B*) Poorly differentiated squamous cell carcinoma. Cancer tissue infiltrated into muscular layer (*A,B:* ×200).

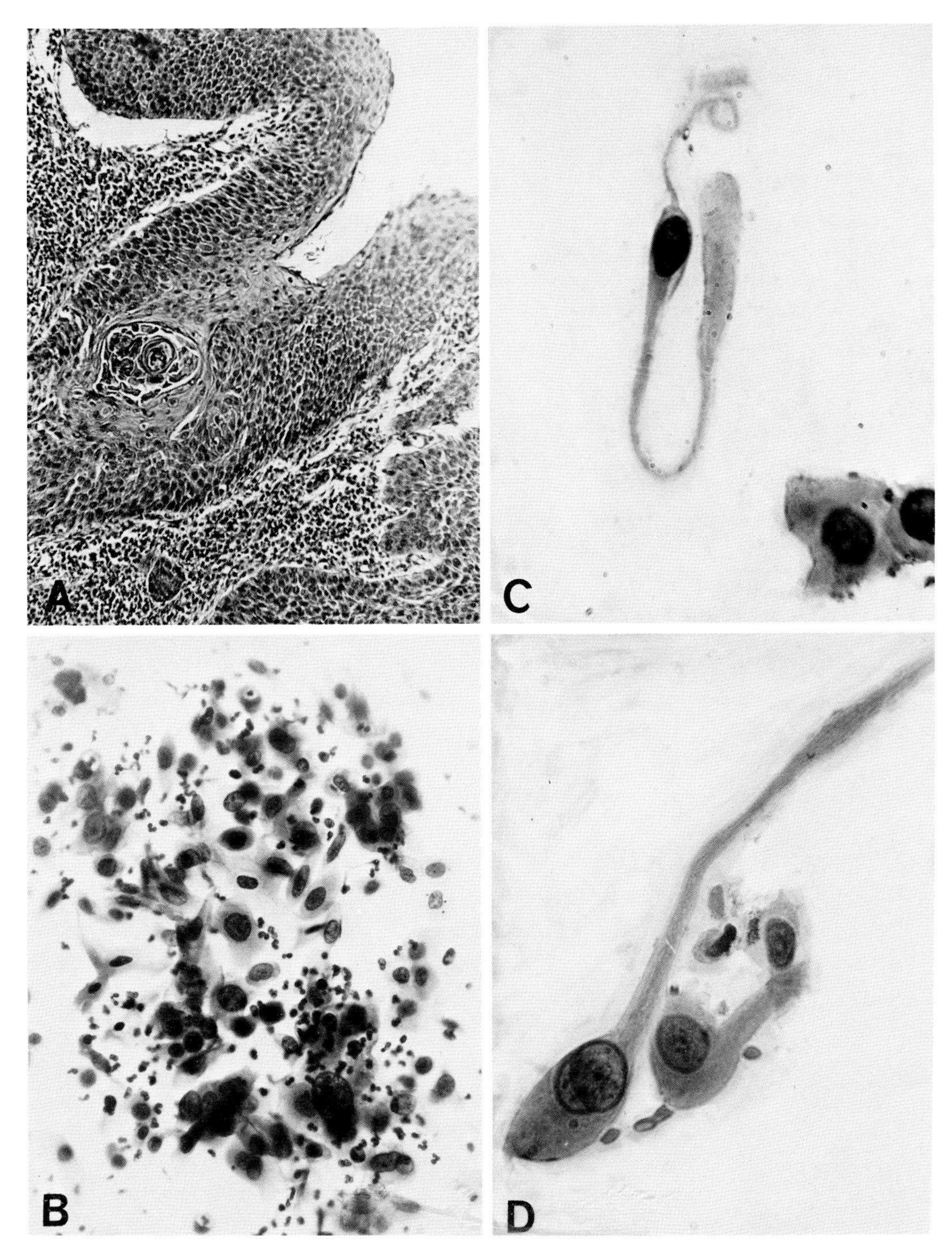

Fig. 7-3. Advanced cancer. (*A*) Well-differentiated squamous cell carcinoma with keratinizing pearl formation (×95). (*B,C,D*) Various types of well-differentiated cancer cells (*B:* ×190; *C,D:* ×380).

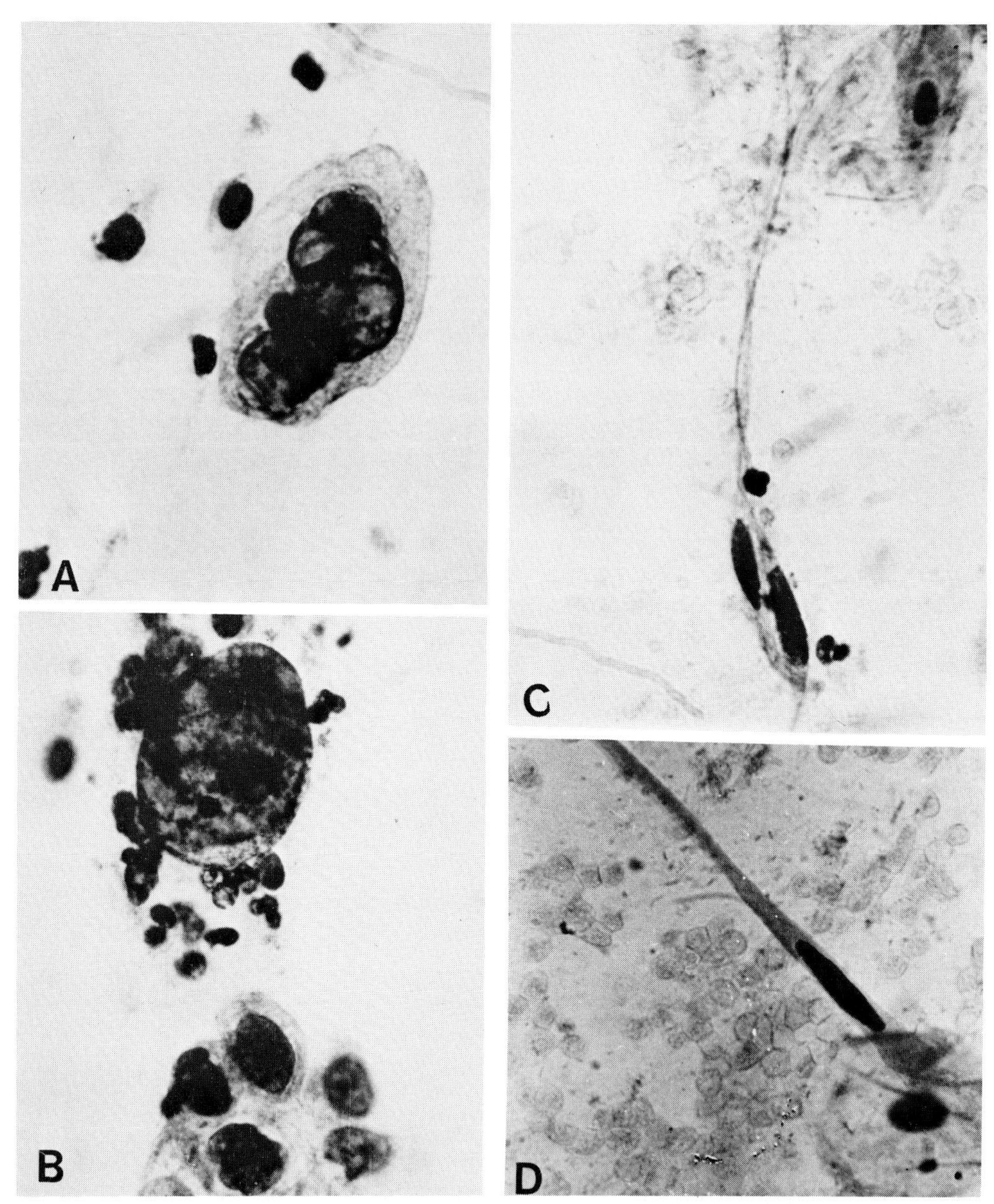

Fig. 7-4. Advanced cancer case. (*A*) Multinucleated cancer cell (×510); (*B*) huge naked nucleus (×510); (*C,D*) fiber cancer cells (×510).

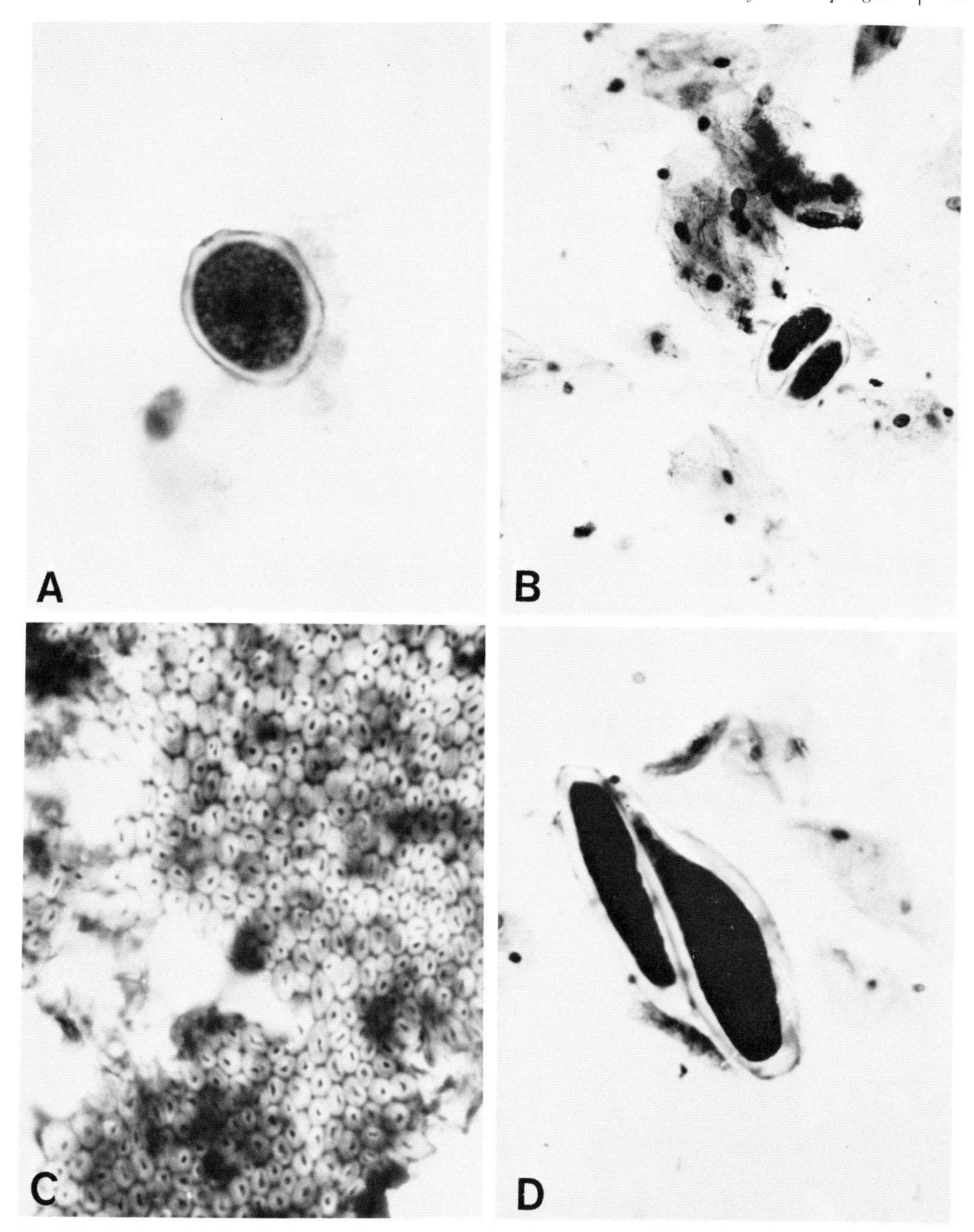

Fig. 7-5. Various kinds of plant cells; pay attention in order to differentiate from carcinoma cells (*A,D:* ×360; *B,C:* ×180).

CHAPTER 8 Sources of Error in the Abrasive Balloon Technique

The balloon technique for the detection of esophageal carcinoma in China is accurate 70–99% of the time. Review of the technique and conversations with its practitioners reveal three possible sources of misdiagnosis.[41]

The major source of error involves the gathering of cells by the balloon or the practitioner's unfamiliarity with the technique. Second, handling the material incorrectly is another cause of misdiagnosis. Misinterpreting the results can also occur.

FALSE-NEGATIVE

The causes for a false-negative result are many. The most common mistake is not having sufficient pressure within the balloon, so that the abrasive net does not come into intimate contact with the wall of the esophagus. Sometimes the practitioner begins with adequate pressure, but upon encountering areas of physiological narrowing, he releases some pressure, but neglects to reinflate the balloon. The ability to determine proper pressure comes with experience. A similar situation also occurs when the balloon's pressure is released before having traversed the full length of the esophagus, thus possibly missing a high lesion.

Often in a patient with an advanced obstructing esophageal lesion, it is impossible to pass the balloon beyond the point of obstruction; thus no cancer cells are obtained and a false-negative reading results. (False-negative rates are higher for advanced stages of the disease.)

If the balloon cannot be passed to 40 cm, which is the expected length of the esophagus, an advanced lesion is suspected and radiographic studies must be performed. Quite often, the reason the balloon cannot be passed is that the lower esophageal sphincter goes into spasms and occludes the lumen. To overcome this problem, the patient is distracted or asked to cough or swallow (but not to drink water, because this would wash the cells downstream).

Another problem facing the inexperienced practitioner is that the balloon and its tubing may coil within the oral cavity. An examination of the patient's mouth will expose the coils of tubing and retrieving the balloon should be no problem. However, the balloon can also work its way into the trachea; violent coughing will be the tip-off. In this case, repeat the procedure 2 or 3 days later.

It is our experience, in examining over 14,002 persons studied by the abrasive balloon method, that 336 (1.65%) of the slides were uninterpretable due to lack of cells on the slide. In our first series of 13,027 cases, 174 cases of esophageal cancer were found (1:75). In our second series of 975 cases, 1 in 50 was the rate of discovery.[2] Moreover, in another study, it was noted that 70% of cases of early carcinoma had fewer than 25 cancer cells on the slide.

The technique for making the smear is another source of false-negative results, especially with respect to early lesions. Mucoid material from the pharynx may prevent detection of esophageal cells by coating the balloon. To avoid this, the balloon is examined grossly for thick, tenacious, mucoid sputum. If present, it is removed by gentle wiping before making the smear. The esophageal sample has a thin, rice-like consistency. Another helpful step to avoid this mucus problem is to ask the patient to clear his throat and to evacuate his nasal cavity before the procedure.

Care must also be taken to ensure that all facets of the balloon are sampled in the smear. Occasionally, a lesion can be missed by inadequate sampling of the balloon.

If the balloon is allowed to rest in the body of the stomach, gastric secretions can coat the net. This makes it difficult for the net to obtain cells as it is pulled up the esophagus. Moreover, even if cells are obtained in spite of gastric secretions, they often float off the slide while being fixed. Gentle blotting of the balloon before making the smear removes the gastric secretions and exposes the esophageal cells for transfer to the slides.

The presence of tissue fragments and/or old clotted blood on the balloon is very suggestive of a lesion; these areas of the balloon must be smeared carefully. Four hundred fifty-six cases were investigated to see whether or not the balloon showed blood. There was blood on the balloon in 89% of the cases that showed cancer. Ninety-three percent of the noncancer cases were without blood on the balloon (see Table 8-1).

To increase the sensitivity of the esophageal balloon technique, smears can be made by washing the balloon in normal saline prior to making a cyto-prep.

The last source of a false-negative result lies in the examination of the smear. In our experience, we have found that both the level of competence of the screener and the speed at which a slide is viewed play an important role (see Table 8-2). Naturally, a situation may arise in which interpretation is difficult for even the most experienced observer. In order to overcome this problem, careful examination must be performed on slides with a suggestive background, i.e., the presence of dysplastic cells, a dirty background, or a suggestive history.

TABLE 8-1
The Relationship between the Cytodiagnosis and the Balloon that Showed Blood

	Cancer		Non-cancer	
Balloon net	No. of cases	%	No. of cases	%
With blood	319	89.6	7	7.0
Without blood	37	10.4	93	93.0
Total	356	100.0	100	100.0

TABLE 8-2
The Results of Re-read Slides in False-Negative Cases

	Re-read slides			
Primary diagnosis	Mild dysplasia	Severe dysplasia	Cancer	Total
Mild dysplasia	31	21	13	65
Severe dysplasia	0	3	4	7
Total	31	24	17	72

FALSE-POSITIVE

Our series yielded fewer false-positive results (including false-suspicious) than false-negative results. Obviously false-positive results were misinterpretations.

Sources for the misinterpretation were: interpreting 1) dysplastic basal cells as squamous cell carcinoma (20 cases); 2) columnar cells from the cardia as adenocarcinoma (3 cases); 3) degenerated naked nuclei as undifferentiated carcinoma (4 cases); and 4) large macrophages, as well-differentiated adenocarcinoma (2 cases).

Because of the great number of patients sampled in our mass surveys in China, slides from numerous patients are commonly processed in one container. "Floaters" (cells or tissue that has floated from one slide to another) have been seen; however, misinterpretation can be avoided by examination of the background, and of course, repeated sampling. This problem should not exist in the United States.

With *careful attention* to the possible sources of misinterpretation, the balloon technique is a highly accurate and reliable method of detection of esophageal carcinoma.

References

1. World Health Organization: World Health Statistics and Causes of Death. WHO, Geneva, 1978.
2. The National Cancer Control Office of Ministry of Health: *Atlas of Cancer Mortality for the People's Republic of China.* China Map Press, Beijing, 1980.
3. Yan, Y.-H.: *Ji Sheng Fang* [medical book of the Sung Dynasty].
4. Wen, H.-R.: *Historical Records of Esophageal Cancer on Ancient Documents of China: Proceeding of Co-ordinating Meeting for Research of Esophageal Cancer in North China.* Shijiazhuang, Hebei Province, 1975.
5. Linxian Research Team for the Prevention and Treatment of Esophageal Cancer: Pharyngo-esophageal and esophageal carcinomas in hens in Linxian of Henan Province. *Acta Zool Sin* **19**(4): 309–342, 1973.
6. Dept. of Pathology of CICAMS: Epidemiological and pathological morphology of pharyngeal and esophageal cancers in domestic fowls. *Acta Zool Sin* **22**(4): 319–326, 1976.
7. Dept. of Pathology of CICAMS and Dept. of Pathology of the Cancer Hospital of Hebei Province: Epidemiology and pathology of pharyngo-esophageal cancers in domestic fowls from Henan immigrant communities and native inhabitants in Zhongxiang County, Hebei Province. *Acta Zool Sin* **22**(4): 314–318, 1976.
8. Dept. of Cancer Epidemiology of Cancer Institute of Chinese Academy of Medical Sciences (CICAMS) et al.: Preliminary investigation of the epidemiological factors of esophageal cancer in China. *Res Cancer Prevent Treat [Zhongliu Fang Zhi Yanjiu]* **2:** 1–8, 1977.
9. Li, M.-H., and Lu, S.-H.: Experimental studies on the carcinogenicity of fungus-contaminated food from Linxian county. In *Genetic and Environmental Factors in Experimental and Human Cancer.* H. V. Gelboin, et al., Eds. Japan Science Social Press, Tokyo, 1980, pp. 139–148.
10. The Coordinating Group for Research on the Etiology of Esophageal Cancer of North China: The epidemiology of esophageal cancer in north China and preliminary results in the investigation of its etiological factors. *Sci Sin* **18**(1): 134–148, 1975.
11. Li, M.-H., and Lu. S.-H.: Formation of carcinogenic N-nitroso compounds in corn-bread inoculated with fungi. *Sci Sin* **22:** 471–477, 1979.
12. Cancer Institute and Ritan Hospital of the Chinese Academy of Medical Sciences: Collection of papers for the twentieth anniversary of the Cancer Institute and Ritan Hospital (a collection of 245 papers, in Chinese), pp. 38–46 and 70–74, Beijing, China, 1978.
13. Shen, C., and Qiu, S. L.: A preliminary report of exfoliative cytology of esophagus. *Chin J Pathol* **7:** 191, 1963.
14. Shu, Y. J., and Chow, B.: Introducing a new instrument for collecting cytological specimen from the esophagus, type 75, 77. *Chin Med J* **56**(9): 571, 1979.
15. Shu, Y. J., and Quiu, S. L.: The five class typing for the epithelial cell in the esophagus. *Chin J Oncol* **1:** 263, 1979.
16. Shu, Y. J., Yany, S. Q., and Gin, S. P.: Further studies on the relationship between esophageal epithelial hyperplasia and esophageal carcinoma. *Nat Med J Chin* **60**(1): 39, 1980.
17. Li, J. Y.: Epidemiology of cancer of the esophagus in China. (To be published.)
18. Shu, Y. J., and Hau, Z. Z.: The correlation of the presence of fungus to the degree of esophageal dysplasia and cancer. *Chin J Oncol* **1:** 2, 1979.
19. Xia, O. J., and Zhan, Y.: Fungal invasion in esophageal tissue and its possible relation to esophageal carcinoma. *Chin Med J* **58**(7): 392–396, 1978.
20. Zhang, H. Z.: Personal communication, 1969.
21. Zhow, Z. C.: The valuation of diagnostic cytology of esophagus. *Chin J Pathol* **9:** 243–245, 1965.
22. Zhen, C., and Oiu, S. L.: A study of esophageal cytology. *Acta Henan Med Coll* **23:** 4–8, 1966.
23. Zhang, H. Z.: Cytologic examination of the esophagus and gastric cardia. A study of 22,805 cases. The conference materials of Linxian County, 1978.
24. (*a*) Shu, Y. J.: Improvement of the apparatus for collecting specimen from esophagus. *Acta Fujian*

Med Coll **2:** 14, 1973. (*b*) Yanting Cancer Team of Sichuan: National reference. (*c*) Chen, S. Z.: National Conference.

25. Cancer Institute, Chinese Academy of Medical Sciences: Improvement of instrument for collecting cytological specimen from esophagus (a study of 600 cases). *Res Cancer Treat Control* **2:** 42, 1975.
26. Panico, F. G., Papanicolaou, G. N., and Cooper, W. A.: Abrasive balloon for exfoliation of gastric cancer cell. *JAMA* **143:** 1308–1311, 1950.
27. Brainsma, A. H.: The value of cytology in the early diagnosis of carcinoma of the esophagus and stomach (making use of the Papanicolaou gastric balloon and its modifications). Thesis, Utrecht, 1957.
28. Sichuan Medical College, Yanting Cancer Team and Jintang People's Hospital: *Res Cancer Treat Control* **3:** 172, 1976.
29. The Coordinating Groups for the Research of Esophageal Carcinoma, Chinese Academy of Medical Sciences and Henan Province: Studies on the relationship between epithelial dysplasia and carcinoma of the esophagus. *Chin Med J* **54:** 679, 1974.
30. Shu, Y. J.: Cytological observation on early esophageal squamous cell carcinoma. A study of 100 cases. *Chin J Oncol* **3:** 39–41, 1981.
31. Shu, Y. J.: An analysis of epithelial dysplasia and carcinoma of the esophagus in progression by cytologic studies. (To be published.)
32. Gai, H. Y., and Shu, Y. J.: Therapy of severe epithelial dysplasia of esophagus: 1. Therapeutic effects of anti-tumor B111, anti-tumor B and tilorone. *Chin J Oncol* **2**(2): 92, 1980.
33. Shu, Y. J.: Morphologic studies of dysplastic lesion in histopathology. (To be published.)
34. Lin, P. Z., and Tsai, H. Y.: *Research of Premalignant Lesion of the Esophagus.* Cancer Institute, Chinese Academy of Medical Sciences annual report, 1978, p. 60.
35. Tumor Prevention, Treatment and Research Group, Chengchow, Henan and Cancer Institute, Chinese Academy of Medical Sciences: Pathology of early esophageal squamous cell carcinoma. *Chin Med J* **3**(3): 180–192, 1977.
36. Lee, T. K.: Cellular measurements on early and advanced squamous cell carcinoma of the esophagus in northern China. *Anal Quant Cytol* **4:** 39–43, 1982.
37. The Coordinating Group of Linxian, Henan Province: Diagnosis and surgical treatment of 170 cases with early esophageal cancer. *Res Esophageal Cancer* **2:** 6, 1975.
38. Shu, Y. J.: Cytologic screening of esophagus in Yao commune of Linxian, Henan Province. *J Prevent Res Esophageal Cancer* **2:** 18–23, 1975.
39. Wu, Y. K.: *The cancer of esophagus and cardia.* Science and Technology, Shanghai, 1964.
40. Wu, X.: *Chinese Academy of Medical Sciences: Collected Papers on Cancer Research.* Publishing House of Science and Technology, Shanghai, 1962, pp. 163–177.
41. Shu, Y. J., Liu, S. F., and He, Z. J.: The factors of cytological misdiagnosis of esophageal cancer and the way of improvement. *Res Cancer Treat Control* **3:** 17, 1979.

Index

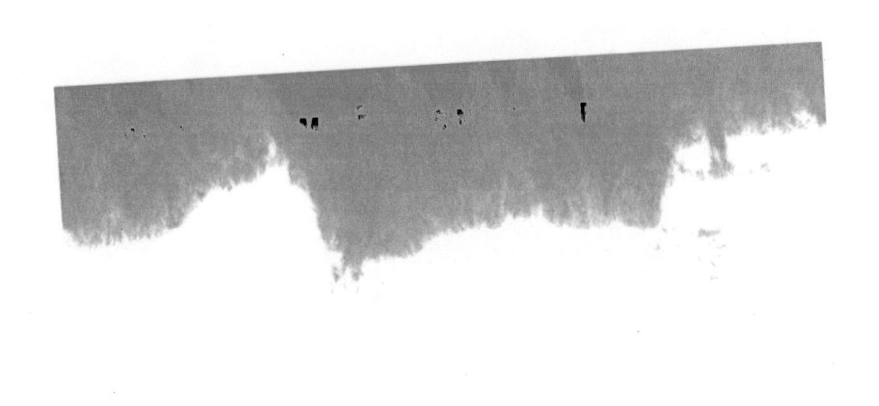